Soil, Plant and Water Analysis Manual

Soil, Plant and Water Analysis Manual

Editor

Arundhati Nimbalkar

scitus
academics

Soil, Plant and Water Analysis Manual

Edited by **Arundhati Nimbalkar**

ISBN: 978-1-68117-098-5

Library of Congress Control Number: 2015935498

SCITUS Academics LLC,
www.scitusacademics.com
Box No. 4766, 616 Corporate Way,
Suite 2, Valley Cottage,
NY 10989

Printed in United States of America on Acid Free Paper ∞

Contents

Preface vii

Chapter 1 **Effects of Hydrogels on Tree Seedling Performance in Temperate Soils before and after Water Stress** 1
Lawrence J. B. Orikiriza, Hillary Agaba, Gerald Eilu, John D. Kabasa, Martin Worbes, and Aloys Hüttermann

Chapter 2 **Adsorption of Diuron and Linuron in Gaza Soils** 21
Yasser El-Nahhal, Mohamed Abadsa, and Samir Affifi

Chapter 3 **An Assessment of Irrigation Water Quality and Selected Soil Parameters at Mutema Irrigation Scheme, Zimbabwe** 37
Abel Chemura, Dumisani Kutywayo, Tapiwanashe M. Chagwesha, and Pardon Chidoko

Chapter 4 **Trace Element Concentration and Speciation in Selected Mining-Contaminated Soils and Water in Willow Creek Floodplain, Colorado** 59
R. Burt, T. Weber, S. Park, S. Yochum, and R. Ferguson

Chapter 5 **Multivariate Analysis of Metal Levels in Paddy Soil, Rice Plants, and Rice Grains: A Case Study from Shakargarh, Pakistan** 109
Saadia R. Tariq and Nabila Rashid

Chapter 6 **Soil Fertility, Phosphorus Fractions, and Maize Yield as Affected by Poultry Manure and Single Superphosphate** 135
A. O. Ojo, M. T. Adetunji, K. A. Okeleye, and C. O. Adejuyigbe

Chapter 7 **Raw Agro-industrial Orange Peel Waste as a Low Cost Effective Inducer for Alkaline Polygalacturonase Production from Bacillus licheniformis SHG10** 157
Amira M Embaby, Aliaa A Masoud, Heba S Marey, Nadia Z Shaban, and Tayssir M Ghonaim

Chapter 8 **Proximate Composition, Mineral Content and Antinutritional Factors of Brebra (Millettia ferruginea) Seed Flour as well as Physicochemical Characterization of its Seed Oil 191**

Berhanu Andualem and Amare Gessesse

Citations..219

Index...223

Preface

Soil is a non-renewable resource upon which mankind depends for survival. Historically, the rise of great civilizations has been linked to the quality of soil and the availability of water. Equally, the demise of such civilizations is often attributed to mismanagement of soil and land in its broadest sense. Crop productivity and soil fertility are thus synonymous. In today's overcrowded world, the challenge to feed and clothe the burgeoning populations of developing countries is a daunting task. Yields have to be increased from existing land areas; adding fertility to the soil to satisfy the demands of higher-yielding crops is essential. Soils vary greatly in their capacity to grow crops without fertilizer; even the richest soils experience declining yields without man's intervention. In essence, soil is not always a perfect medium for growing plants; it is, however, the only one that is available. As no essential element can substitute for another, it is critically important to identify where and when such deficiencies occur. That's where the role of soil and plant analysis comes in. Techniques have been developed to evaluate soil fertility constraints based on soil chemical extraction and analysis of the plants that grow on such soils. Both are complementary and, when calibrated with field crop responses to fertilizer, provide a rational basis to identify what elements are missing, and how much fertilizer, whether organic or inorganic, to apply. Therefore, soil and plant analysis laboratories have a vital role in agricultural development of the WANA region. However, the process does not end there. To be meaningful and valid, tests have to be appropriate for the purpose intended and reliable and repeatable.

Editor

Chapter 1

Effects of Hydrogels on Tree Seedling Performance in Temperate Soils before and after Water Stress

Lawrence J. B. Orikiriza[1], Hillary Agaba[2], Gerald Eilu[1], John D. Kabasa[3], Martin Worbes[4], and Aloys Hüttermann[5]

[1]School of Forestry, Environmental and Geographical Sciences, Makerere University, Kampala, Uganda

[2]National Forestry Resources Research Institute, Kampala, Uganda

[3]Department of Biosecurity, Ecosystems and Veterinary Public Health, Makerere University, Kampala, Uganda

[4]Department for Crop Sciences, Institute of Tropical Agriculture, University of Goettingen, Goettingen, Germany

[5]Faculty of Forest Sciences and Forest Ecology, University of Goettingen, Goettingen, Germany

ABSTRACT

Super Absorbent Polyacrylate (SAP) hydrogels absorb and store water thereby aiding plant establishment when incurporated in the soil. The effect of cross-linked SAP hydrogel amendment on the performance of tree seedlings of Picea abies, Pinus sylivestris and Fagus sylvatica grown in temperate soils under water stress and non-water stress periods was investigated in a green house. The objective was to compare the root and shoot biomass of seedlings of the three species grown in sand, loam and clay soils amended with 0.4% w/w hydrogel in non-water stress conditions as well as survival, root and shoot biomass after subjection to water stress. The seedlings were grown for 16 weeks, harvested and shoot as well as root biomass determined before water stress. The seedlings were also subjected to water stress and their biomass assessed at death following the water stress. The results showed that root and shoot biomass were generally higher in hydrogel amended soils compared to the controls. Root and shoot biomass of Fagus sylvatica was lower compared to Picea abies and Pinus sylivestris before water stress. The 0.4% hydrogel amendment significantly increased species' survival in the different soils studied. Although root biomass was higher in hydrogel amended sandy soil compared to other soils, P. sylivestris and F. sylvatica shoot biomass were higher in hydrogel amended clay and loam soils compared to the sandy soil after water stress. Biomass was higher in sand compared to loam and clay soils under non-water and water stressed conditions. Since SAP hydrogel amendment improved the survival and biomass production of tree seedlings before and after water stress, use of SAPs could be promoted to enhance seedling production in water stress and non-water stress environments.

INTRODUCTION

Water scarcity is a challenge to agriculture and plantation forestry in many parts of the tropical and temperate regions in the world. Water saving technologies that enhance plant establishment and growth in soils of different properties are required. Soils generally differ in moisture content, temperature and mineralogy [1] which may require different soil moisture conservation technologies. One available technology is the use of super absorbent hydrophilic polymers [2]. Super absorbent polymers (SAPs) are substances that can retain large quantities of water and nutrients when incorporated in the soil, making it available for plant growth whenever required. They can be linear or cross-linked hydrogels [3] based on the structure of the cross-linking agents. This study focused on the latter type that has a relatively higher water absorption capacity compared to the former. The soil water and nutrients stored in SAPs are released gradually for plant growth under water limiting conditions [4, 5] whereas under non-water limiting conditions, they are reported to enhance nutrient uptake for plant growth [3].

Different soils and tree species exhibit varying responses to SAP hydrogel amendment. Several studies have shown that addition of hydrogels to growing media increased water holding capacity by up to 400% [6] and decreased water stress by delaying the onset of wilting

[7]. in water-stressed soils for instance, prolonged survival and improved growth have been recorded under drought conditions [8-12] and in sandy desert soils [13]. SAPs were reported to increase tree species survival and reduce evapo-transpiration in different soils under drought conditions [14]. Improved growth has also been reported in seedlings grown in well watered sandy soil [15] as well as tropical soils [16].

Since the effects of SAP hydrogels on trees grown in temperate soils and specifically under non-water stress conditions are not known, it is necessary to understand the behaviour of SAPs in a wide range of soils in order to recommend which trees for what type of soils. Most studies on hydrogel application have been conducted on sandy soils [17-22] with less attention to loamy and clayey soils. Thus, the behavior of hydrogels in the latter soils remains largely unexplored and could potentially limit their application over a wide range of agro-ecological zones. This study therefore investigated tree seedlings performance in temperate soils before and after water stress. The objective of the study was to compare survival and biomass of seedlings of the following tree species (Picea abies (L.) (H. Karst.) (Spruce), Pinus sylivestris L. (Pine) and Fagus sylvatica L. (Beech) in temperate soils amended with SAPs before and after water stress. The following hypotheses were tested: 1) Hydrogel amendment increases the biomass of tree seedlings in sandy, loamy and clayey soils before water stress, 2) Hydrogel amendment prolongs the survival time of tree seedlings in sandy, loamy and clayey soils exposed to drought and 3) There is no difference in tree biomass between hydrogel amended soils before and after subjecttion to water stress.

MATERIALS AND METHODS

Soils

Sand, loam and clay soils were used in the study. The sand was obtained from a sand pit in Schöningen (Longitude 9°40›0» E Latitude: 51°38›0» N) in the Solling Mountains close to Göttingen, Germany. The loam soil was sampled from the field of the Institute for Tropical Agriculture, Georg August University Göttingen (Longitude 9°56′2″ E Latitude 51°32′1″ N). The clay soil was collected from the forest at the clay factory in Göttingen. Samples of the soils were collected from 30 cm deep 10 m × 10 m pits and separately mixed into one composite sample each of about 2 kg at the Fakultät für Forstwissenschaften und Waldökosysteme, Soil science laboratory, Georg August University.

All samples were air dried, sieved through a 2 mm sieve, oven dried at 80°C for 72 hours and their chemical characteristics analyzed. A pH meter was used to determine the soil pH [23] while organic C and N were determined by the Walkley-Black [24] and Kjeldahl [25] methods. Available phosphorus was analyzed using the Bray method [26,24]. Exchangeable cations (Ca^{2+}, Mg^{2+}, $K^{+}$and Na^{+}) were extracted by shaking the soil sample for 2 hours with 1 M Ammonium acetate [27]. Concentrations of K and Na were then determined by a flame photometer whereas Ca and Mg were determined by Atomic Absorption Spectrophotometry [26]. The characteristics of the soils are presented in Table 1.

Plant Material

Six month old seedlings of Picea abies (Spruce), Pinus sylivestris (Pine) and Fagus sylvatica (Beech) obtained from a commercial tree nursery in Göttingen Germany were used in the experiment. The seedlings were healthy and free from pests and diseases. These species were chosen because they are widespread in Europe, are found in a large range of ecological conditions and have economic value. Picea abies for example, is one of the most common and economically important coniferous species in Europe and Scandinavia and tolerates acidic soils although it is not well suited for dry or nutrient deficient soils

[28]. Pinus sylvestris is native to Europe and Asia. It is an important plantation forestry tree that is used to reforest degraded coal mines and burned sites. Seedlings of Pinus sylvestris establish best in soils with adequate moisture and some shade. Survival is best when the seedlings are planted on microsites close to the tops of hills, and lowest in overly moist depressions. Fagus sylvatica, the European Beech or Common Beech, is a deciduous tree that grows well in almost any type of soil. However, it grows best in fairly humid areas with well-drained soils found on moderately fertile ground, calcified or lightly acidic.

Hydrogel

Luquasorb hydrogel manufactured by the BASF SE Chemical Company, Ludwigshafen, Germany was used to amend the different soils at 0.4% hydrogel concentration. The control had no hydrogel added. The 0.4% hydrogel concentration was made by mixing 4 kg of hydrogel powder with 1000 kg of soil prepared in a concrete mixer (Mini Concrete Mixer, Model: CM 180-MZ2; Mixing capacity 180 L). The amount of hydrogel and the mixing procedure followed previous studies [4] and recommendations by the manufacturer.

Table 1: Chemical characteristics of the soils used in the experiment

Parameter	Sand	Loam	Clay
pH	4.37	6.20	5.80
Carbon (%)	0.06	0.70	0.40
Nitrogen (%)	0.01	0.21	0.12
Available Phosphorus (ppm)	9.81	14.80	10.14
Calcium (Meq/100g soil)	1.47	3.00	1.82
Magnesium (Meq/100g soil)	0.33	0.93	1.03
Potassium (Meq/100g soil)	0.08	0.32	0.25
Sodium (Meq/lOOg soil)	0.06	0.12	0.15

Treatments and Experimental Design

Thirty five seedlings each of Picea abies, Pinus sylivestris and Fagus sylvatica were transplanted into 3 kg polythene pots each filled with sand, loam and clay soils amended at either 0.4% hydrogel or a control (no hydrogel amendment). Altogether this made 18 treatment combinations in a Randomized Factorial Design i.e. (Sand + hydrogel × 3 tree species, Sand + control × 3 tree species; Loam + hydrogel × 3 tree species, Loam + control × 3 tree species; Clay + hydrogel × 3 tree species, Clay + control × 3 tree species). Each of the 18 treatment combinations of seedlings in hydrogel amended or UN amended soils were randomly placed in a greenhouse and maintained at 25°C - 32°C and relative humidity of 50% - 95%. Depending on whether or not the soils were saturated, the plants were watered for at least once a day for 16 weeks until they were established. Proper establishment was indicated by the growth of new leaves or needles and twigs.

Determination of Biomass before Water Stress

After 16 weeks, the seedlings were watered to field capacity (when the water settled on the soil surface and drainage became negligible). This ensured complete hydrogel expansion and soil saturation. Ten seedlings in each treatment combination were harvested and cut into two parts-roots and shoots, oven dried for 72 hours, then weighed using a sensitive Sartorius weighing scale (Model ED 8201-CW Extend Precision balance 8200 × 0.1 g) to give the dry weight before water stress. The roots were separated from the soil by gently washing under slow moving clean tap water.

Determination of Survival and Biomass after Water Stress

The remaining 25 were subjected to water stress by complete termination of watering and monitored daily to observe initiation of wilting, desiccation and death. The start date of desiccation was recorded (T1). The date when a seedling died was recorded (T2). The seedlings were monitored until the color of the stems, leaves, branches and/or needles changed from green to brown or grey. A plant was recorded as dead when all the leaves and stems turned brown and started falling off as branches became brittle. Brittleness was ascertained by breaking a sample branch which fell off confirming the brittleness. Roots and shoots were also oven dried for 72 hours, weighed using a sensitive Sartorius weighing scale (Model ED 8201-CW Extend Precision balance 8200 × 0.1 g) to obtain the dry weight after water stress.

Data Analysis

One-way Analysis of variance (ANOVA) was used to test the effect of the factors (hydrogel and soil types) on survival time and biomass production each tree species at $p < 0.05$ in SPSS version 16. The mean root and shoot biomass across hydrogel levels were compared between soil types using paired t-tests. Biomass before and after desiccation was also compared using paired t-tests. All tests were carried out at $p < 0.05$ level of significance.

RESULTS AND DISCUSSIONS

Effect of Hydrogel on Biomass in Non-Water Stressed Soils

Root and shoot biomass of the three species were generally higher in hydrogel amended soils compared to the controls. However, Fagus sylvatica root and shoot biomass were generally lower than for Picea abies and Pinus sylivestris in hydrogel amended soils (Figure 1). Overall, root biomass was generally higher in hydrogel amended sandy soil compared to other soils. In the case of P. sylivestris and F. sylvatica, however, shoot biomass was higher in hydrogel amended clay and loam soils compared to the sandy soil (Figure 1).

These findings confirm that SAP hydrogel amendments improve tree growth performance in non-water stressed temperate soils. Hydrogel induced biomass responses between species and soils in this study concur with those obtained from green house experiments on tropical soils in Uganda [16]. Differences in the species' biomass accumulation responses to hydrogel amendment for instance, the generally lower biomass of Fagus sylvatica relative to Picea abies and Pinus sylivestris could be attributed to the different soil moisture requirements of tree species caused by genotypic differences between the species. Considering that soils with higher Cation Exchange Capacity (CEC) have the capacity to retain plant nutrients better [29], the increase in biomass when water is not limiting is possibly a result of improved nutrient availability to the plants, following SAP hydrogel application. It is possible that incorporation of hydrogels to soils provided additional adsorption of cations to the negative charges thereby increasing the soil base saturation hence enhancing biomass.

The relatively high root biomass in hydrogel amended sandy soil compared to other soils is related to the low water retention capacity of loam and clay soils. Divalent cations in soils (e.g. Ca^{2+} and Mg^{2+}) can destroy the polymer lattice of hydrogels thus diminishing water retention by the gel [18]. Hydration of gels in the presence of divalent (Ca^{2+} and Mg^{2+}) and monovalent (K^{+} and NH^{4+} at 20 meq. per liter) cations reduces from 10% and 20% mM [30]. The relatively high levels of these elements in loam and clay soils relative to sand (Table 1) therefore partly explain the observed biomass differences. However, the effect of calcium and magnesium cations on water absorption by loam and clay soils was not specifically investigated.

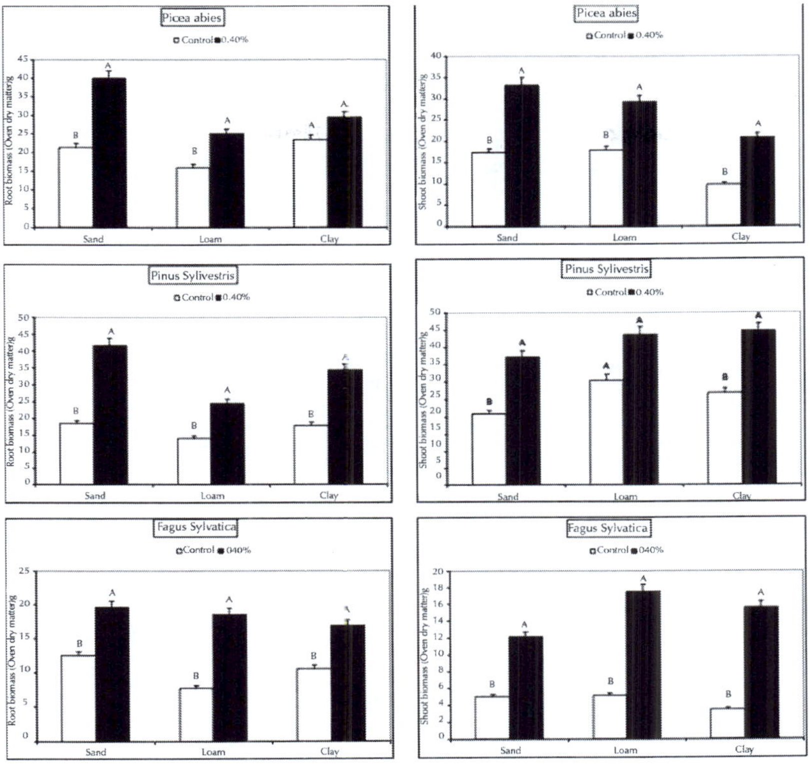

Figure 1: Root and shoot biomass responses of tree seedlings of Picea abies. Pinus sylivestris and Fagus sylvatica to hydrogel amendment in different soils before water stress. Each value is the mean of 5 plants. Different letters in the same bar cluster show significant differences.

Hydrogel Effect on Seedling Survival and Biomass in Water Stressed Soils

Survival time of Picea abies, Pinus sylivestris and Fagus sylvatica seedlings significantly increased in the 0.4% hydrogel amended sand, loam and clay soils compared to the controls (Figure 2). In hydrogel amended soils survival was 66, 71 and 57 days more in sand, 8, 11 and 3 days more in loam and 17, 3 and 4 days more in clay soils respectively for Picea abies, Pinus sylivestris and Fagus sylvatica seedlings compared to the controls. Species' root and shoot biomass was generally higher in hydrogel amended soils compared to the controls (Figure 3).

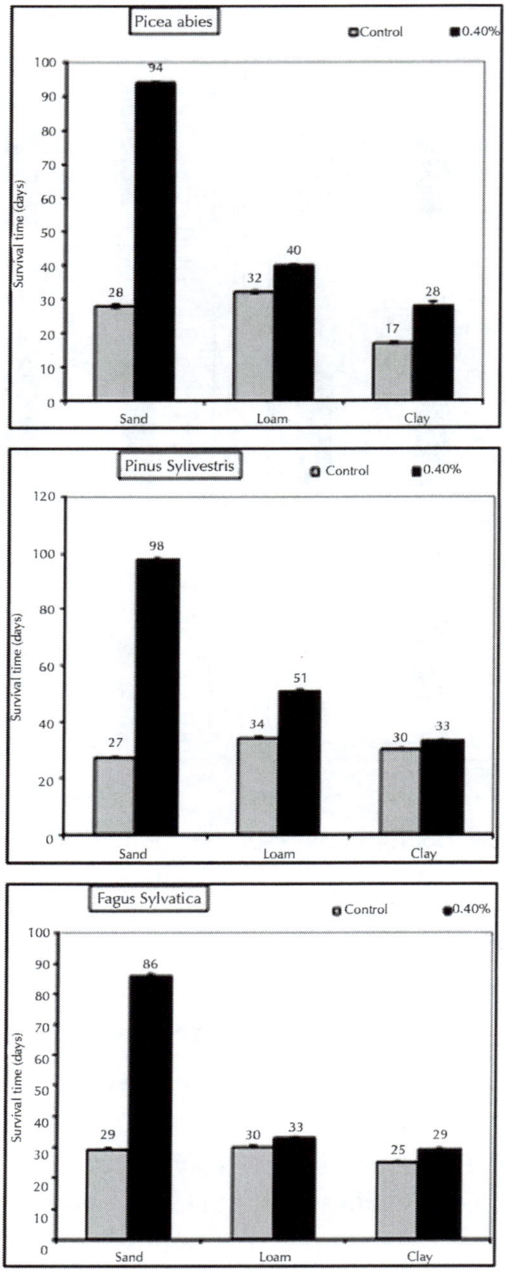

Figure 2: Survival time after water stress of selected tree seedlings grown in different soils amended with hydrogel. Each value is the mean of 5 plants. Error bars indicate standard error of the mean.

The higher duration of species' seedling survival, root and shoot biomass responses in hydrogel amended soils compared to the controls, confirm that SAP hydrogel amendment improves survival and tree growth performance in water stressed temperate soils. Hydrogels enhance the water holding capacity of soils and thus provide supplementary plant available water to plant root zones in dry soils [13, 31]. Whereas biomass was higher in hydrogel amended soils compared to the controls, the effect was higher in sand compared to loam and clay (i.e. overall, hydrogel had a marginal effect on root biomass in clay and loam soils). Our results on improved survival time, root and shoot biomass growth during water stress conditions agree with previous studies in different media, [11,32,33,34]. Despite the differences (e.g. in water retention) between hydrogel amended temperate soils [35], the relatively higher root and shoot biomass responses for P. abies and P. sylivestris compared to F. sylvatica may be explained by the genotypic differences between tree species [36]. Earlier studies (e.g. [18,36] report similar different plant responses to hydrogel treatments in drought stressed soils). The magnitude of increase in biomass following hydrogel amendment may be attributed to the differences in strategies of species in responding to water stress [37,38]. Although prolonged water stress inhibits plant growth through its effects on physiological processes such as CO_2 assimilation [11] hydrogel amendment improves the photosynthetic rate, root growth, reducing CO_2 assimilation and stomatal conductance inhibition caused by water stress [11]. It is thus likely that the species used in this experiment, strategically partitioned resources in response to water stress; with for instance Picea abies and Pinus sylivestris dramatically developing more root and shoot biomass, whereas Fagus sylvatica reduced root biomass. This is especially beneficial in dry environments where more root growth implies increased capacity to absorb the scarce water resources whereas increased shoot growth means a larger leaf area for photosynthesis with more growth [39] and enhanced water use efficiency. Hydrogel amended sandy soils have a higher water retention capacity [2,14]. It is also possible that large pore spaces in sandy soils allow more swelling of the hydrogel thereby providing more supplementary water available for plant growth compared to the loam and clay soils.

Change in Biomass during Water Stress

Total biomass (root and shoot) in hydrogel amended soils ranged from 5 to 45 times higher in hydrogel amended soils when compared to the controls after subjecting trees to water stress (Figure 4). However, total biomass gen-erally reduced following water stress although some increases were recorded in hydrogel amended sand and loam soils (Figure 4).

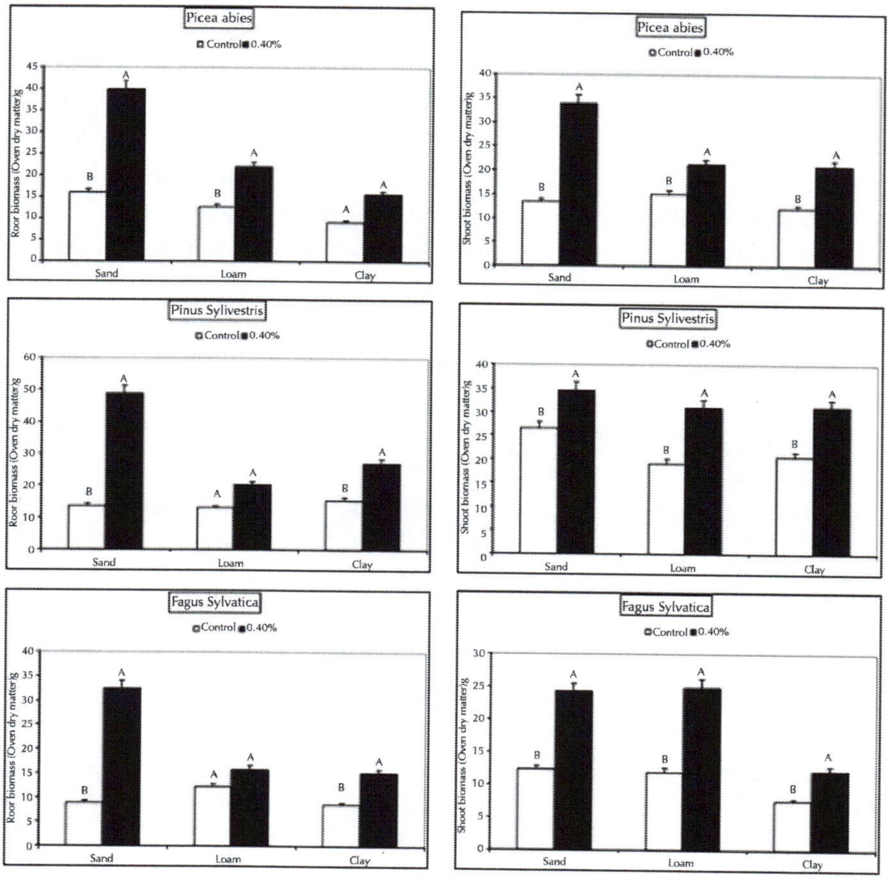

Figure 3: Responses of different tree species to hydrogel amendment during water stress conditions. Each value is the mean of 5 plants. Different letters in the same bar cluster are significantly different.

Changes in total biomass for the species in hydrogel amended soils are signals of physiological adjustments to water stress. This partly accounts for the observed biomass increases in sand and loam soils amended with hydrogel during desiccation. Reduced biomass in hydrogel amended soils appears to be related to the ability of hydrogels to delay drought stress effects [33]. However, the fact that hydrogels enhanced total biomass in different soils before and after desiccation, implies that hydrogels can be applied in soils with different moisture contents (i.e. from dry to saturated at field capacity).

CONCLUSIONS

In conclusion, hydrogel amendment increased tree seedling root and shoot biomass of Picea abies, Pinus sylivestris and Fagus sylvatica seedlings in sand, loam and clay soils in non-water stressed soils. Under water stressed conditions, hydrogel amendment prolonged the species' survival and improved biomass in sandy soils compared to other soils. Species' biomass generally reduced the following water stress. Such effects of SAPs can be used to promote tree seedling production and planting programmes especially in water stress and non-water stress environments.

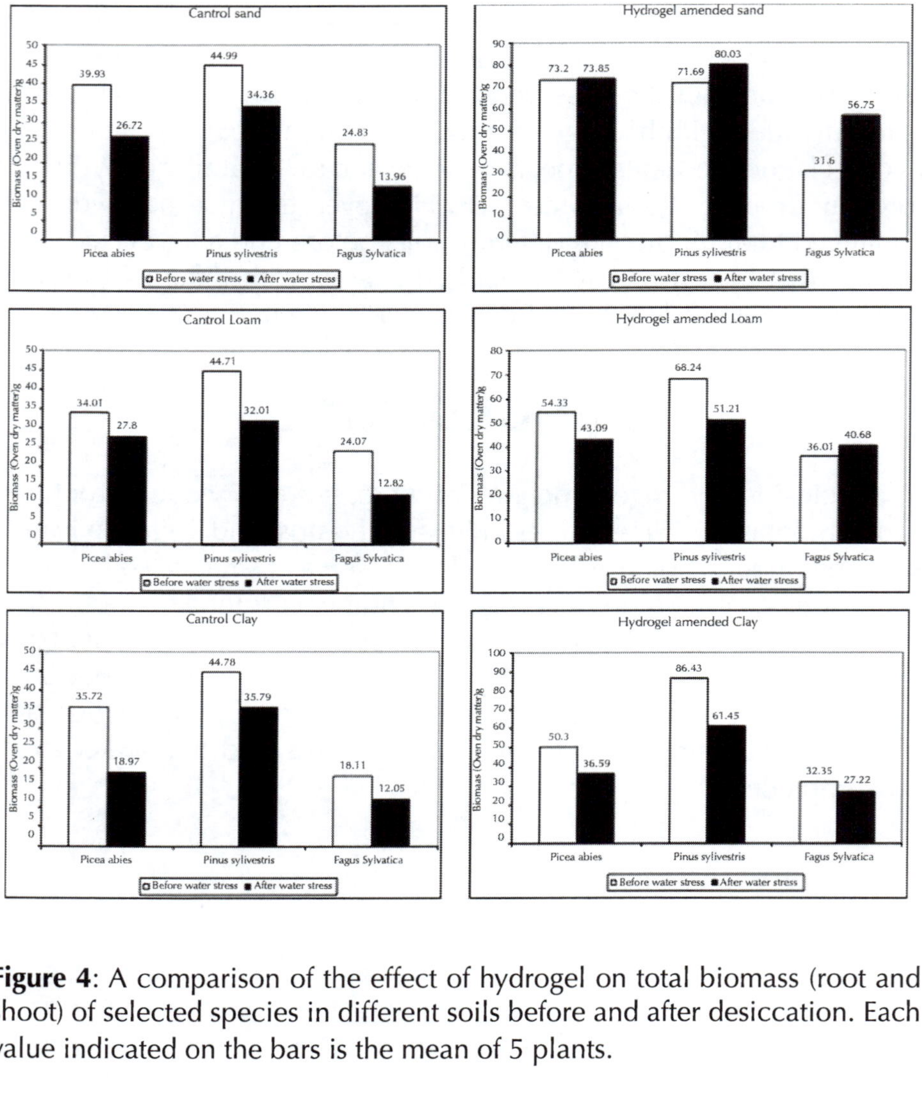

Figure 4: A comparison of the effect of hydrogel on total biomass (root and shoot) of selected species in different soils before and after desiccation. Each value indicated on the bars is the mean of 5 plants.

ACKNOWLEDGEMENTS

We are grateful to the KAAD and BASF SE, Chemical Company, Germany, for jointly funding this study and providing the stipends of Lawrence Orikiriza and Hillary Agaba. We also pay a special tribute to Prof. Dr. Aloys Hüttermann who passed on during the preparation of this work. We dedicate this work to his honor and memory.

REFERENCES

1. J. Six, C. Feller, K. Denef, S. M. Ogle, J. C. Moraes Sa and A. Albrecht, "Soil Organic Matter, Biota and Aggregation in Temperate and Tropical Soils—Effects of NoTillage," Agronomie, Vol. 22, No. 7-8, 2002, pp. 755-775. doi:10.1051/agro:2002043
2. S. S. Dorraji, A. Golchin and S. Ahmadi, "The Effects of Hydrophilic Polymer and Soil Salinity on Corn Growth in Sandy and Loamy Soils," Clean-Soil, Air, Water, Vol. 38, No. 7, 2010, pp. 584-591.
3. A. Hüttermann, L. J. B. Orikiriza and H. Agaba, "Application of Superabsorbent Polymers for Improving the Ecological Chemistry of Degraded or Polluted Lands," Clean-Soil, Air, Water, Vol. 37, No. 7, 2009, pp. 517-526. doi:10.1002/clen.200900048
4. A. Hüttermann, M. Zommorodi and K. Reise, "Addition of Hydrogels to Soil for Prolonging the Survival of Pinus halepensis Seedlings Subjected to Drought," Soil & Tillage Research, Vol. 50, No. 3-4, 1999, pp. 295-304. doi:10.1016/S0167-1987(99)00023-9
5. F. Yazdani, I. Allahdadi and G. A. Akbari, "Impact of Superabsorbent Polymer on Yield and Growth Analysis of Soybean (Glycine max L.) under Drought Stress Condition," Pakistan Journal of Biological Sciences, Vol. 10, No. 23, 2007, pp. 4190-4196.doi:10.3923/pjbs.2007.4190.4196
6. M. S. Johnson, "Effects of Soluble Salts on Water Absorption by Gel-Forming Soil Conditioners," Journal of the Science of Food and Agriculture, Vol. 35, No. 10, 1984, pp. 1063-1066. doi:10.1002/jsfa.2740351004
7. J. M. Gehring and A. J. Lewis, "Effects of Hydrogel on Wilting and Moisture Stress of Bedding Plants," Journal of the American Society for Horticultural Science, Vol. 105, No. 4, 1980, pp. 511-513.
8. M. Sarvas, "Effect of Desiccation on the Root System of Norway Spruce (Picea abies [L.] Karst.) Seedlings and a Possibility of Using Hydrogel STOCKOSORB for Its Protection," Journal of Forest Science, Vol. 49, No. 11, 2003, pp. 531-536.
9. E. M. A. Youssef, "Influence of Hydrogels Compounds on Acclimatization Behavior of Acacia Melanoxylon. Br. Vitroplants," Acta Horticulturae, Vol. 608, 2003, pp. 81- 84.

10. J. Akhter, K. Mahmood, K. A. Malik, A. Mardan, M. Ahmad and M. M. Iqbal, "Effects of Hydrogel Amendment on Water Storage of Sandy Loam and Loam Soils and Seedling Growth of Barley, Wheat and Chickpea," Plant, Soil and Environment, Vol. 50, No. 10, 2004, pp. 463-469.

11. V. Arbona, J. D. Iglesias, J. Jacas, E. Primo-Millo, M. Talon and A. Gomez-Cadenas, "Hydrogel Substrate Amendment Alleviates Drought Effects on Young Citrus Plants," Plant and Soil, Vol. 270, No. 1, 2005, pp. 73-82. doi:10.1007/s11104-004-1160-0

12. K. G. Apostol, D. F. Jacobs and R. K. Dumroese, "Root Desiccation and Drought Stress Responses of Bare Root Quercus rubra Seedlings Treated with a Hydrophilic Polymer Root Dip," Plant and Soil, Vol. 315, No. 1-2, 2009, pp. 229-240. doi:10.1007/s11104-008-9746-6

13. J. Abedi-Koupai, S. S. Eslamian and J. A. Kazemi, "Enhancing the Available Water Content in Unsaturated Soil Zone Using Hydrogel, to Improve Plant Growth Indices," Ecohydrology and Hydrobiology, Vol. 8, No. 1, 2008, pp. 67-75.

14. H. Agaba, L. J. B. Orikiriza, J. F. O. Esegu, J. Obua, J. D. Kabasa and A. Hüttermann, "Effect of Hydrogel Amendment to Different Soils on Plant Available Water and Survival of Trees under Drought Conditions," Clean-Soil, Air, Water, Vol. 38, No. 4, 2010, pp. 328-335. doi:10.1002/clen.200900245

15. J. P. Syvertsen and J. M. Dunlop, "Hydrophilic Gel amendments to Sand Soil Can Increase Growth and Nitrogen Uptake Efficiency of Citrus Seedlings," Horticultural Science, Vol. 39, 2004, pp. 267-271.

16. L. J. B. Orikiriza, H. Agaba, G. Eilu, M. Tweheyo, J. D. Kabasa and A. Hüttermann, "Amending Soils Hydrogels Increases the Biomass of Nine Tree Species under NonWater Stress Conditions," Clean-Soil, Air, Water, Vol. 37, No. 8, 2009, pp. 615-620.doi:10.1002/clen.200900128

17. K. C. Taylor and R. G. Halfacre, "The Effect of Hydrophilic Polymer on Media Water Retention and Nutrient Availability to Ligustrum lucidum," Horticultural Science, Vol. 21, 1986, pp. 1159-1161.

18. J. M. Woodhouse and M. S. Johnson, "The Effect of a Gel Forming Polymer on Seed Germination and Establishment," Journal of Arid Environments, Vol. 20, 1991, pp. 375-380.

19. M. Silberbush, E. Adar and Y. De-Malach, "Use of an Hydrophilic Polymer to Improve Water Storage and Availability to Crops Grown in Sand Dunes I. Corn Irrigated by Trickling," Agricultural Water Management, Vol. 23, No. 4, 1993, pp. 303-313. doi:10.1016/0378-3774(93)90042-9

20. A. M. Al-Omran and A. R. Al-Harbi, "Improvement of Sandy Soils with Soil Conditioners," In: A. Wallace and R. E. Terry, Eds., Handbook of Soil Conditioners: Substances That Enhance the Physical Properties of Soil, Marcel Dekker, Inc., New York, 1997, pp. 363-384.

21. A. I. Al-Humaid and A. E. Moftah, "Effects of Hydrophilic Polymer on the Survival of Buttonwood Seedlings Grown under Drought Stress," Journal of Plant Nutrition, Vol. 30, No. 1, 2007, pp. 53-66. doi:10.1080/01904160601054973

22. A. K. Bhardwaj, I. Shainberg, D. Goldstein, D. N. Warrington, G. C. Levy, "Water Retention and Hydraulic Conductivity of Cross-Linked Polyacrylamides in Sandy Soils," Soil Science Society of America, Vol. 71, No. 2, 2007, pp. 406-412.doi:10.2136/sssaj2006.0138

23. J. M. Anderson and J. S. I. Ingram, "Tropical Soil Biology and Fertility: A Hand-Book of Methods," CAB International, Wallingford, 1993.

24. D. W. Nelson and L. E. Sommers, "Total Carbon, Organic C and Organic Matter," In: A. L. Page, et al., Eds., Methods of Soil Analysis, Part 2, Agronomy, 2nd Edition, ASA, Madison, 1982, pp. 539-579.

25. J. M. Bremner, "Total N," In: C. A. Black, et al., Eds., Methods of Soil Analysis, Part 2 Agronomy, ASA, Madison, 1965, pp. 1149-1178.

26. R. H. Bray and L. K. Kurtz, "Determination of Total Organic and Available Forms of Phosphorus in Soils," Soil Science, Vol. 59, No. 1, 1945, pp. 39-45.doi:10.1097/00010694-194501000-00006

27. J. R. Okalebo, K. W. Gathua and P. L. Woomer, "Laboratory Methods of Soil and Plant Analysis Manual," TSBF-CIAT and SACRED Africa, Nairobi, 2000.

28. G. Vogel Willis, "A Guide for Revegetating Coal Mine Soils in the Eastern United States," Department of Agriculture, Forest Service, Northeastern Forest Experiment Station, 1981, 190 p.
29. K. F. Wenger, "Forestry Hand Book," John Wiley & Sons Inc., Hoboken, 1984.
30. D. C. Bowman, R. Y. Evans and J. L. Paul, "Fertilizer salts Reduce Hydration of Polyacrylamide Gels and affect Physical Properties of Gel-Amended Container Media," Journal of American Society of Horticultural Science, Vol. 115, No. 3, 1990, pp. 382-386.
31. D. L. Bouranis, A. G. Theodoropoulus and J. B. Drossopoulus, "Designing Synthetic Polymers as Soil Conditioners," Communications in Soil Science and Plant Analysis, Vol. 26, No. 9-10, 1995, pp. 1455-1480. doi:10.1080/00103629509369384
32. Z.-B. Luo, K. Li, X. Jiang and A. Polle, "Ectomycorrhizal fungus (Paxillus involutus) and Hydrogels Affect Performance of Populus euphratica Exposed to Drought Stress," Annals of Forest Science, Vol. 66, No. 1, 2009, p. 106. doi:10.1051/forest:2008073
33. Y. Shi, J. Li., J. Shao, S. Deng, R. Wang, N. Li, J. Sun, H. Zhang, H. Zhu, Y. Zhang, X. Zheng, D. Zhou, A. Hüttermann and S. Chen, "Effects of Stockosorb and Luquasorb Polymers on Salt and Drought Tolerance of Populus popularis," Scientia Horticulturae, Vol. 124, 2010, pp. 268-273.
34. J. Abedi-Koupai and J. Asadkazemi, "Effects of a Hydrophilic Polymer on the Field Performance of an Ornamental Plant (Cupressus arizonica) under Reduced Irrigation Regimes," Iranian Polymer Journal, Vol. 15, No. 9, 2006, pp. 715-725. doi:10.1016/j.scienta.2009.12.031
35. J. L. Moya, F. R. Tadeo, A. Gómez-Cadenas, E. PrimoMillo and M. Talón, "Transmissible salt Tolerance Traits Identified through Reciprocal Grafts between Sensitive Carrizo and Tolerant Cleopatra Citrus Genotypes," Journal of Plant Physiology, Vol. 159, No. 9, 2002, pp. 991- 998. doi:10.1078/0176-1617-00728
36. M. Ahmad and H. Verplancke, "Germination and Biomass Production as Affected by Salinity in Hydrogel Treated Sandy Soil," Pakistan Journal of Forestry, Vol. 44, No. 2, 1994, 53-61.
37. P. J. Kramer and J. S. Boyer, "Water Relations of Plants and Soils," Academic Press, San Diego, 1995.

38. B. Duan, Y. Lu, C. Yin, O. Junttila and C. Li, "Physiological Responses to Drought and Shade in Two Contrasting Picea asperata Populations," Physiologia Plantarum, Vol. 124, No. 4, 2005, pp. 476-484. doi:10.1111/j.1399-3054.2005.00535.x

39. R. Van-den-Boogaard, D. Alewijnse, E. J. Veneklaas and H. Lambers, "Growth and Water-Use Efficiency of 10 Triticum aestivum Cultivars at Different Water Availability in Relation to Allocation of Biomass," Plant, Cell and Environment, Vol. 20, No. 2, 1997, pp. 200-210. doi:10.1046/j.1365-3040.1997.d01-60.x

Chapter 2

Adsorption of Diuron and Linuron in Gaza Soils

Yasser El-Nahhal, Mohamed Abadsa, and Samir Affifi

Department of Earth and Environmental Sciences, the Islamic University of Gaza, Gaza, Palestine

ABSTRACT

This study investigated the adsorption potential of diuron and linuron in selected soil samples collected from agricultural area of Gaza Strip, Palestine and correlated the adsorption results to the soil properties. The adsorption experiments are based on the batch equilibrium technique and UV-Spectrophotometer method to determine the equilibrium concentration of both herbicides. Results of adsorption experiment showed that linuron and diuron are adsorbed in various amounts in Gaza soils according to the variation in pH, organic matter and clay content. Adsorption isotherms of both herbicides were linear in the tested concentration. Fitting the adsorption data to Freundlich equation showed good fitting and Freundlich constants were less than one indicating physical adsorption processes. It is concluded that soil organic matter and pH affect the adsorption of both herbicides. These results provided a better understanding of the behavior of diuron and linuron in Gaza soils.

INTRODUCTION

Diuron and Linuron are phenylurea herbicides widely used in the world to control annual weeds [1]. The amount used in Gaza Strip Palestine, progressively increased in the past ten years [2]. This situation may have led to groundwater contamination. However, recent studies indicate toxicity to cyanobacteria [3], and Fish [4]. Several trials have been made to reduce environmental contamination [5-7]. These studies are based on adsorbing the herbicides to organo-clay complexes. This process restricts the leaching potential of pesticides. However, it is still important to understand the adsorption behavior of herbicides in soil. Chiristen et al. [8] and Nkedi-Kizza et al. [9] studied the adsorption behavior of phenylurea herbicides in different soils and reported that organic carbon was the main factor affecting urea sorption. Chaplain et al. [10] reported that soil hydrophobicity was an important factor for diuron adsorption in soils. Whereas, several researchers [11-13] reported that the adsorption of linuron was significantly correlated with soil organic matter and clay content. Haouari et al. [12] found that the adsorption of diuron and linuron in clayey soils in Morocco fitted the Freundlich equation. Furthermore, Yihua Liu et al. [14] studied the adsorption-desorption behaviors of diuron investigated in six cultivated soils of China and reported that diuron adsorption on soil was at a rather high level under low pH value conditions and decreased with increasing pH value. Moreover, Jianhua et al. [15] studied the adsorption kinetics of diuron from aqueous solutions onto activated carbon fiber, and showed the formation of hydrogen bonding between diuron and water, and temperature variations may possibly affect the adsorption process. This study investigated the adsorption behavior of linuron and diuron in selected soils from Gaza Governorates.

MATERIALS AND METHODS

Technical Diuron and Linuron (purity = 99%) were purchased from Sigma-Aldrich company in Germany. Physicochemical properties of the tested compounds are shown in Table 1. Technical Diuron and Linuron (purity = 99%) were purchased from Sigma-Aldrich company in Germany.

Soil Sampling

The soil samples were collected from top soils (0 - 30 cm) of three agricultural locations in Gaza governorates (Khanyounis Governorate, Middle Governorate and North Gaza Governorate). The samples were transferred to a plastic bag, air-dried, sieved through 2 mm, and stored in well-closed plastic bottles in the laboratory. Site description, identification data and coordinates were recorded as shown in the Table 2.

Determination of Soil Texture

The soil texture expresses the proportions of the various size classes (clay < 0.002 mm, silt 0.002 - 0.02 mm and sand 0.02 - 2.0 mm particle size). The proportions of these fractions were determined by Hydrometer method using ASTM 152-H hydrometer [16].

Determination of Soil pH

Soil pH values were measured potentiometrically in a 1: 2.5 soil-water suspension. Twenty grams of a driedsieved soil were transferred into 200 ml beaker. 50 ml of distilled water were added while stirring for one hour using Electric magnetic stirrer. The pH meter was calibrated using pH buffer 4.0, 7.0 and 9.0, and then the pH of suspension was measured according to previous report [17].

Electrical Conductivity (EC)

Electrical conductivity (EC) of soil was measured potentiometrically in a 1: 2.5 soil-water suspensions. Twenty grams of an oven dried-sieved soil were weighed, transferred into 200 ml beaker. 50 ml of distilled water were added while stirring for one hour using Electric magnetic stirrer. The Electrical conductivity meter was calibrated using Standard potassium chloride (KCl) solutions 0.01 and 0.1 M and then the EC of filtrate were measured, Unit of measurement dS/m [18].

Determination of the Organic Matter

The organic carbon content in the soils used were analyzed in Ministry of Agriculture in Gaza using Walkley-Black method [19]. The organic carbon in the sample is oxidized with potassium dichromate and sulfuric acid. The excess potassium dichromate is titrated against ferrous ammonium sulfate. One gram soil was weighed and transferred into 500 mL conical flask. 10 mL of 1 N $K_2Cr_2O_7$ and 20 mL of conc. H_2SO_4 were added. Swirled carefully then let to stand for 30 minutes. 200 mL distilled water and 10 mL H_3PO_4 were slowly added. Then 1 mL of diphenylamine indicator was added and the resulted suspension was titrated against 0.5 N ferrous ammonium sulfate solution until green color started to appear indicating the end point. Blank must run simultaneously.

Adsorption Experiments

Stock solution of diuron/linuron was prepared by dissolving 30 mg active ingredient in 2 - 3 mL methanol and diluting to 1 L with deionised water. The low concentration of methanol in the adsorption experiments had no influence on herbicide adsorption [20]. The adsorption of diuron/ linuron on soil was measured at room temp. (25°C ± 2°C). Appropriate aliquots of the aqueous stock solution of a diuron/linuron was diluted with distilled water to 25 mL and added to 50 mg soil in 30-mL centrifuge tubes. The concentration of diuron/ linuron ranged between 1.2 mg/L and 31 mg/L. The final concentration of soil was 1 g/L. The dispersions were kept under continuous agitation during 48 hours. The supernatant was separated by centrifugation at 20,000 g for 0.5 h.

Table 1: Physicochemical properties of the tested compounds

Physical Properties	Dimon Linuron	Physical Properties
Chemical Name	N-(3.4-dichloropheny1)-N, N-dimethyl urea	3-(3,4-dichloropheny1)-1-methoxy -l-methylurea
Molecular Weight	233.10	249.11
Water Solubility	42 mg/L	81 mg/L
Vapor Pressure	0.41 mPa	2 mPa
Partition Coefficient (Kow)	2.6	3.0043

Table 2: Sampling sites identification information

Site	City	Prominent Mark	GPS Coordinates	
			N	E
1	ICIL Y	Alsouraje Street	31,22,150	34,20,656
2	Middle zone	Elberka Street	31,24,905	34,20,243
3	North Zone	Safe Agricultural Society	31,22,102	34,31,493

The concentration of herbicide in the supernatants was determined by UV-spectrophotometer, (CT-220 Spectrophotometer), the wavelengths of the absorbance of diuron are 247 nm and for linuron is 246 nm according to previous report [21].

Linear regression was used to determine the equilibrium concentration of herbicide solutions. The regression showed R^2 value close to unity (0.9992). The amount of linuron, diuron adsorbed was calculated from the depletion of the linuron concentration by adsorption according to Equation (3.2) El-Nahhal and Safi [22].

$$C_iV = C_eV + MS \tag{3.1}$$

$$S = (C_i - C_e)V / M \quad (3.2)$$

where C_i is the initial concentration of herbicide, and C_e is the remaining concentration of the herbicide in the solution in mg/L, V is the volume of the solution in litter, S is the concentration of the herbicide in the solid phase mg/g (the adsorbed amount), and M is the mass of soil in gram.

For each isotherm, a reference solution with an intermediate concentration was stirred without soil to evaluate adsorption on the glass or other losses. All adsorption experiments were made in duplicate samples with a control.

Standard Curve

Stock solution of diuron and linuron was diluted in 1ml of grade methanol and then diluted in water to a concentration of 38 ppm diuron and 35.4 ppm of linuron as a working standard. A series of linuron standards of, 0.00, 0.56, 2.26, 7.08, 14.16 and 21.24 ppm were prepared. The absorbance was measured spectrophotometrically at 246 nm. A series of diuron standards of 0.00, 0.506, 1.518, 3.036, 12.144, and 24.288 ppm, were prepared. The absorbance measured spectrophotometrically at 247 nm.

Statistical Analysis

The growth inhibition data was analyzed for variance, and main effects and interactions was tested for significance using repeated measures ANOVA. Univariate comparisons of mean growth inhibition at different depths were performed by T-test (a = 0.05), the statistical analysis was performed by using Microsoft Excel software.

RESULTS AND DISCUSSION

Soil Properties

Properties of soil used in this study are shown in Table 3. It can be seen that pH values of the tested soils range between 7.32 and 8. Furthermore the EC values range from 1.81 and 2.63 dS/m, the total organic matter range from 0.801% and 0.254%. This indicates that the soil is nearly poor with organic matters. The clay fraction of soils range between 10% and 47%. The Characteristics of used soil are shown in Table 3.

Adsorption Isotherm of Linuron

The relationship between the Optical Density (OD) and low concentration of Diuron and linuron showed linear relationship (Data not shown) with R^2 values of 0.9997 and 0.997 diuron and linuron respectively. These results indicate strong positive association. Accordingly, the regression equations of linear relationship of Diuron (Y = 0.082X) and linuron (Y = 0.0827X) were used to determine the remaining concentration of the corresponding compound in the aqueous solution in the adsorption experiment. Y and X represent OD and the remaining concentration respectively. Results of the control of adsorption experiments indicate no changes in the initial concentration of linuron and diuron in the glass tube that did not contain soil in the adsorption. Explanation of these results suggests that the glass tube used for adsorption experiment has no capacity to adsorb the herbicides. Accordingly, the disappearance of linuron or diuron in the experiment is due to the adsorption in the soil fraction. EL-Nahhal and Safi, [22] found similar results for other cases.

Adsorption of diuron

Figure 1 presented the plot of adsorption of diuron in various soils from Gaza Governorates; it appears that the adsorbed amount in North Gaza (NZ) Governorate soil is the highest among the selected soils.

Table 3: Soils characteristics

Site	pH	EC dS/m	OM%	TOC%	Sand%	Silt%	Clay%	Tex
Kh. Y	8	1.59	0.76	0.44	62.5	17.5	20	SL
MZ	7.32	1.81	0.25	0.15	87.5	1.25	11.3	LS
NZ	7.33	2.36	0.57	0.33	60	12.5	27.5	C
Clay soil	7.36	2.63	0.80	0.47	23	30	47	CS

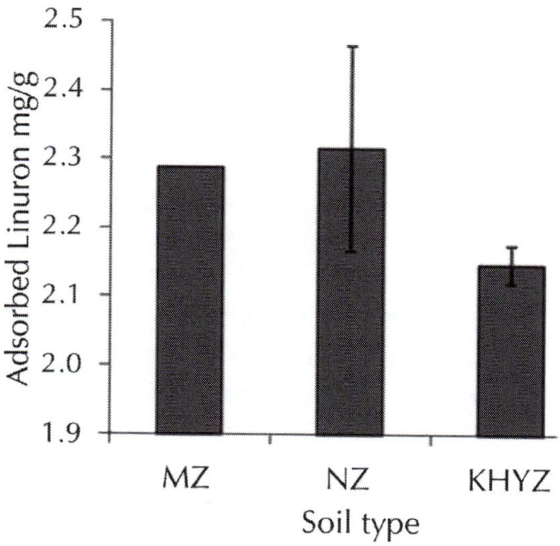

Figure 1: Adsorption of linuron in different soils. Error bars represent standard deviation.

This can be explained by the fact that NZ soil has the highest clay fractions (27.3%) and organic matter (0.571%), followed by Khanyounis Governorate (KHYZ) soil which has the second highest clay mineral content (20%) and has the highest organic matter content 0.762% (Table 3).

The adsorbed amounts in NZ and KHYZ soil are higher than that of Middle Zone (MZ). This probably is due to the high organic matter and clay fraction (Table 3), this probably is due to cation exchange capacity in the clay.

Clay and organic matter are often intimately associated; a previous report found that the organic matter was the key adsorbent for diuron ([23,24]. Afifi and Abu-Swareh, [25] found that the organic matter in Gaza soils play an important role in the adsorption of chemicals. Adsorption isotherm of diuron in Kh.Y soil is shown in Figure 2.

It can be seen that the adsorption of diuron increases as its concentration in the equilibrium solution is increased, indicating strong positive correlation.

Adsorption of Linuron

Adsorptions of linuron in various soils from Gaza are presented in Figure 3. It appears that the adsorbed amount in NZ, and Kh.Y soil is higher than that of MZ, this probably is due to the high organic matter and clay fraction (Table 3). A previous report found that adsorption of linuron was significantly correlated with soil organic matter [11, 13]. The soil of Kh.Y and NZ have high fraction of organic carbon 0.442% and 0.331% respectively (Table 3). This indicates that organic carbon was the main factor affecting linuron adsorption. Similar observation was previously reported [8]. Adsorption of linuron in the North Governorate and in the Khanyounis are nearly similar degrades less to the high value of error bars. The explanation of this results is that both soils have close pH values making the ionization fraction are similar.

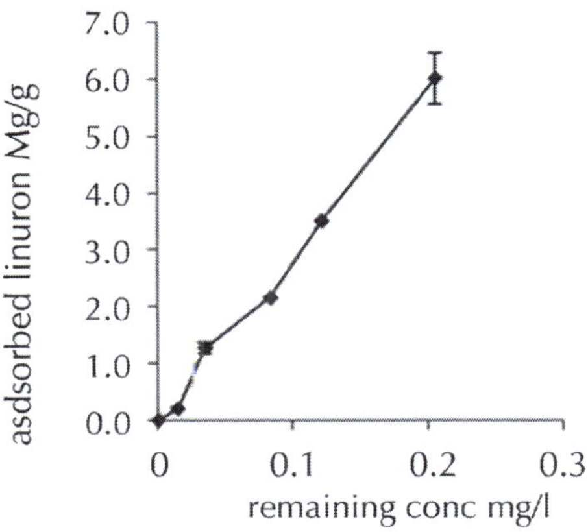

Figure 2: Adsorption isotherm of linuron on Kh.Y soil. Error bars represent standard deviation.

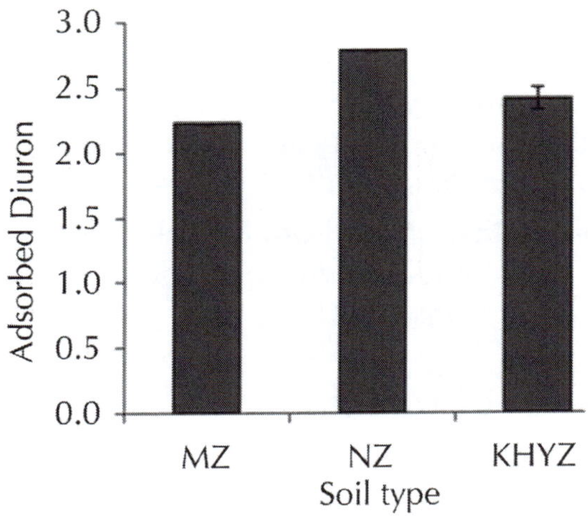

Figure 3: Adsorption of diuron on different soils in Gaza.

Accordingly, similar amount of adsorbed linuron were observed. Adsorption isotherm of linuron in Kh.Y soil is shown in Figure 4. The adsorbed amount increased with increasing the concentration of linuron. Application of linear regression appears best fit in the relationship between points and R^2value equals to 0.99 indicating strong positive association. This suggests high affinity of these soil samples for linuron at low herbicide concentrations. It appears that the adsorbed amount is very little (0.6%) the explanation of this result is that linuron has hydrophobic nature and the soil is more hydrophilic due to the presence of organic cations in soil (Table 3), a previous report found same results in other case [21]. Nevertheless, the adsorbed amount of linuron increased linearly as the concentration increased, probably due to the high presence of clay fraction in soil (Table 3). EL-Nahhal & Lagely [20] found similar phenomena for the adsorption of linuron in Bentonite clays, Sorensen et al., [13] found that the linuron is bound to soil (especially clay) and organic matter and does not move freely.

Classification of Adsorption Isotherms

Adsorption isotherm of Diuron (Figure 2) and linuron (Figure 4) can be classified as L-type according to the classification of Giles et al., [26]. It was stated that the magnitude of the exponent 1/n gives an indication of the favorability and capacity of the adsorbent/adsorbate system. The soils of Kh.Y showed L-type isotherm (1/n > 1), which indicates that Diuron/linuron is adsorbed on the clay fraction of soil, besides organic matter. Our results agreed with previously [10,23] who produced similar results with other soils. Furthermore, the low adsorption of diuron in Kh. Y soil (Figure 2) is probably due to high pH in Kh.Y soil (Table 3). It may be suggested that at high pH value Diuron becomes an anion, in this case repulsion forces between clay surfaces and diuron reduced the adsorption. Under this condition, the organic fraction in soil is the main adsorbent. At low pH value Diuron may become a proton and higher adsorbed amount may be observed This statement is strongly supported by ElNahhal and Safi [22] and Yihua liu et al., [14] who found that the adsorption of diuron on the selected soils were rather high at low pH values and decreased with the increasing pH values of the suspension.

Fitting the data of Diuron (Figure 1) and linuron (Figure 4) to Freundlich equation show linear relationship with R^2 = 0.9978 and 0.9873 respectively. Both herbicides have low K and n values (Table 4). These results indicate physical interaction between Diuron/linuron and clay or soil organic matter fraction.

The low value of n (<1) indicates physical adsorption and monolayer adsorption.

The isotherm is of type L according to the previous classification [26], which shows the adsorption curve for diuron in clay soil at relatively low range of equilibrium concentration. This low concentrations range is prompted by the low solubility of diuron, the adsorption curve for clay soil show initial slight increase in adsorption with increase in concentration.

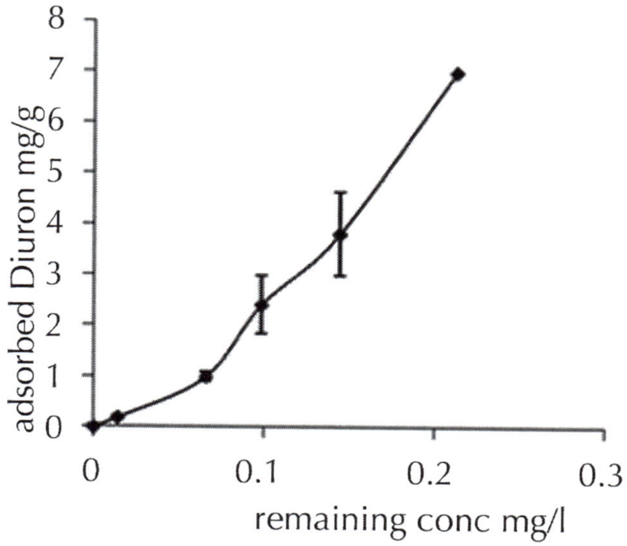

Figure 4: Adsorption isotherm of dinuron on Kh.Y soil. Error bars represent standard deviation.

Table 4: Freundlich constants for diuron and linuron for Kh. Y soil

Herbicide	Khanyounis soil		
	K	n	r2
Dimon Linuron	49.2 50.82	0.752 0.77	0.986 0.997

Comparing the adsorption of Diuron and linuron in Kh.Y soils, it can be seen that Diuron is adsorbed more than linuron. This may be due to the hydrophobic factor. It is obvious that diuron has lower Kow value than linuron (Table 1) indicating less hydrophobic tendency. A number of researchers have hypothesized additional site-specific interactions between the polar groups of the organic compounds and charges on the mineral surface [27].

CONCLUSIONS

This study presents an attempt to understand the adsorption potential of Diuron and linuron in selected soil from Gaza Governorate, Palestine. Our idea in this study provides a better understanding to the behavior of herbicides in Gaza Soils. The study is based on the following approach:

- The first step is to collect soil samples from different agricultural area and to characterize the physicochemical properties of soil and to correlate them with the adsorption results.
- The second step is to understand the adsorption behavior using batch equilibrium technique and spectrophotometer to determine the equilibrium concentrations of the tested herbicides.

The adsorbed amount of Diuron and linuron is moderate in all soils. The adsorption isotherms of both herbicides were best fit to Freundlich equation with r^2 values equal to 0.99 for both herbicides. The physical adsorption parameter was less than "one" for all cases. Generally the adsorption of those herbicides was poor in most soils and considerably affected by three factors:

- Clay fraction is negatively charged and exchange cations are present, they can influence the adsorption of ionic molecules by ion exchange.
- Soil pH.
- Soil organic matter which acts as a non-polar phase or surface and consequently is the main sorbent for diuron and linuron.

ACKNOWLEDGEMENTS

Dr Y. El-Nahhal acknowledges Alexander von Humboldt Stiftung/ Foundation Fellowship Grant no IV-PAL/ 1104842 STP, Germany.

REFERENCES

1. C. D. Tomlin, "The Pesticides Manual," 11th Edition, Britich Crop Protection Council, 2000.

2. Ministry of Agriculture, "Annual Report," Palestinian National Authority, Ram Allah, PNA, 2010.
3. S. Kerkez, "Eco-Toxicity of Diuron, Dequat and Terbutryn as Single and Mixtures to Cyanobacterial Mats," M.Sc. Thesis, The Islamic University of Gaza, Gaza, 2013.
4. Sh. EL-Njjar, "Eco-Toxicity of Carbaryl, Chlorpyrifos, and Diuron, as Individual and in Mixtures on Daphnia magna and Tilapia nilotica," MSc Thesis, The Islamic University-Gaza, Gaza, 2013.
5. Y. El-Nahhal, S. Nir, S. Serban, O. Rabinowitz and B. Rubin, "Organoclay Formulation of Acetochlor for Reduced Movement in Soil," Agricultural and Food Chemistry, Vol. 49, No. 11, 2001, pp. 5371-5464. doi:10.1021/jf010561p
6. G. Lagaly, "Introduction: Pesticide-Clay Interactions and Formulations," Applied Clay Science, Vol. 18, No. 5-6, 2001, pp. 205-209. doi:10.1016/S0169-1317(01)00043-6
7. G. Rytwo, Y. Gonen, S. Afuta and S. Dultz, "Interaction of Pendimethalin with Organo-Montmorillonite Complexes," Applied Clay Science, Vol. 28, No. 1-4, 2005, pp. 67-77. doi:10.1016/j.clay.2004.01.016
8. C. M. Fouquk-Brouard and J. M Fournier, "AdsorptionDesorption and Leaching of Phenylurea Herbicideson Soils," Talanta, Vol. 43, No. 10, 1996, pp. 1793-1802.
9. P. Nkedi-Kizza, P. C. Rao and J. W. Johnson, "Adsorption of Diuron and 2,4,5-T on Soil Particle-Size Separates," Environmental Quality, Vol. 12, No. 2, 1983, pp. 195-197.doi:10.2134/jeq1983.00472425001200020009x
10. V. Chaplain, A. Brault, D. Tessier and P. Defossez, "Soil Hydrophobicity: A Contribution of Diuron Sorption Experiments," Soil Science, Vol. 59, No. 6, 2008, pp. 1202- 1208.doi:10.1111/j.1365-2389.2008.01080.x
11. M. Sánchez-Camazano, M. J. Sánchez-Martín and R. Delgado-Pascual, "Adsorption and Mobility of Linuron in Soils as Influenced by Soil Properties, Organic Amendments, and Surfactants," Agriculture and Food Chemistry, Vol. 48, No. 7, 2000, pp. 3018-3026.doi:10.1021/jf990812i
12. H. Haouari, A. Dahchour, A. Peña-Heras, X. Louchard, B. Lennartz, M. Elbelghiti Alaoui and A. Satrallah, "Behavior of Two

Phenylurea Herbicides in Clayey Soils and Effect of Alternating Dry-Wet Conditions on Their Availability," Environmental Science and Health, Part B, Vol. 41, No. 6, 2006, pp. 883-893.

13. S. R. Sorensen, J. Rasmussen, C. S. Jacobsen, O. S. Jacobsen, R. K. Juhler and J. Aamand, "Elucidating the Key Member of a Linuron-Mineralizing Bacterial Community by PCR and Reverse Transcription-PCR Denaturing Gradient Gel Electrophoresis 16s rRNA Gene Fingerprinting and Cultivation," Applied and Environmental Microbiology, Vol. 71, No. 7, 2005, pp. 4144-4148. doi:10.1128/AEM.71.7.4144-4148.2005

14. Y. H. Liu, Z. Z. Xu, X. G. Wu, W. J. Gui and G. N. Zhu, "Adsorption and Desorption Behavior of Herbicide Diuron on Various Chinese Cultivated Soils," Hazardous Materials, Vol. 178, No. 1-3, 2010, pp. 462-468.

15. J. H. Xu, Y. B. Sun, Z. Y. Li and J. W. Feng, "Kinetic of Diuron Adsorption on to Activated Carbon Carbon Fiber Study," School of Environment, Vol. 2, 2011, pp. 832- 835.

16. G. J. Bouyoucos, "Hydrometer Method Improved for Making Particle Size Analyses of Soils," Agronomy Journal, Vol. 54, No. 5, 1962, pp. 464-465.doi:10.2134/agronj1962.0002196200540 0050028x

17. S. Dhyan, P. K. Chhonkar and R. N. Pandey, "Soil, Plant and Water Analysis—Amethod Manual," IARI, New Delhi, 1999.

18. L. P. Van Reeuwijk, "Procedures for Soil Analysis," 3rd Edition, International Soil Reference and Information Centre (ISRIC), Wageningen, 1992.

19. J. T. Sims and A. M. Wolf, "Recommended Soil Testing Procedures for the Northeastern United States," Northeastern Regional Publication No. 493 (Revised), University of Delaware Agricultural Experimental Station, Newark, 1995.

20. Y. El-Nahhal and G. Lagaly, "Salt Effects on the Adsorption of a Pesticide on Modified Bentonites," Colloid and Polymer Science, Vol. 283, No. 9, 2005, pp. 968-974.

21. I. Franco, C. Vischetti, M. T. Baca, M. De Nobili, C. Mondini and L. Leita, "Adsorption of Linuron and Metamitron on Soil and Peats at Two Different Decomposition Stages," Soil and Sediment Contamination: An International Journal, Vol. 6, No. 3, 1997, pp. 307-315.

22. Y. El-Nahhal and J. Safi, "Adsorption of Benzene and Naphthalene to Modified Montmorillonite," Journal of Food, Agriculture and Environment, Vol. 3, No. 2, 2005, pp. 295-298.

23. A. G. Ahangar, R. J. Smernik, R. S. Kookana and D. J. Chittleborough, "Clear Effects of Soil Organic Matter Chemistry, as Determined by NMR Spectroscopy, on the Sorption of Diuron," Chemosphere, Vol. 70, No. 7, 2008, pp. 1153-1160. doi:10.1016/j.chemosphere.2007.08.054

24. P. Wang and A. A. Keller, "Sorption and Desorption of Atrazine and Diuron onto Water Dispersible Soil Primary Size Fractions," Water Research, Vol. 43, No. 5, 2009, pp. 1448-1456. doi:10.1016/j.watres.2008.12.031

25. S. Afifi and A. Abu-Swareh, "Adsorption Behaviour of 2,4-Dichlorophenol and 2,4,5-Trichlorophenol on Soil of Gaza Strip," Bethlehem University, Vol. 18, 1999, pp. 36-54.

26. C. H. Giles, T. H. Mac Ewan, S. N. Nakhwa, and D. A. Smith, "System of Classification of Solution Adsorption Isotherms, and Its Use in Diagnosis of Adsorption Mechanisms and in Measurement of Specific Surface Areas of Solids," Chemical Society, Vol. 111, 1960, pp. 3973- 3993. doi:10.1039/jr9600003973

27. G. Y. Sheng, Y. N. Yang, M. S. Huang and K. Yang, "Influence of pH on Pesticide Sorption by Soil Containing Wheat Residue-Derived Char," Environmental Pollution, Vol. 134, No. 3, 2005, pp. 457-463. doi:10.1016/j.envpol.2004.09.009

Chapter 3

An Assessment of Irrigation Water Quality and Selected Soil Parameters at Mutema Irrigation Scheme, Zimbabwe

Abel Chemura[1], Dumisani Kutywayo[2], Tapiwanashe M. Chagwesha[3], and Pardon Chidoko[4]

[1]Department of Environmental Science & Technology, Chinhoyi University of Technology, Chinhoyi, Zimbabwe

[2]DR & SS, Agricultural Research Centre, Harare, Zimbabwe

[3]AGRITEX, Chipinge, Zimbabwe

[4]DR & SS, Coffee Research Institute, Chipinge, Zimbabwe

ABSTRACT

Soil salinity and sodicity are major factors limiting agricultural productivity in irrigation schemes located in semi-arid areas. A study was conducted to assess the quality of irrigation water used in Mutema

Irrigation Scheme located in south-east Zimbabwe to understand how irrigation water quality is related to the chemical quality of soils in the scheme. Irrigation water samples were collected from groundwater and surface sources in 2012 and their hydrochemistry determined while soil samples were collected from irrigated and non-irrigated parts of the scheme in 2006 and 2012 and analysed for selected chemical properties. The results indicated that the groundwater had high concentrations of Na^+ (4.35 mg/l), Mg^{2+} (4.75 mg/l), Cl^- (3.6 mg/l) and Electrical Conductivity (EC) (1729 Msm/cm) compared to the surface irrigation water source which had 0.72 mg/l Na^+, 2.25 mg/l Mg^{2+}, 0.78 mg/l Cl^- and 594 Msm/cm EC. The soils in the scheme had higher levels of pH, Sodium Adsorption Ratio (SAR), Exchangeable Sodium Percentage (ESP) and EC which in some blocks exceeded the threshold requirements for cropping. It was found that pH, SAR and ESP were significantly higher ($p < 0.05$) in irrigated blocks compared to non-irrigated areas of the scheme, indicating an influence of irrigation water on soils characteristics in irrigated plots· Mg^{2+} and Ca^{2+} in the soils positively correlated with Na+ ($r^2 = 0.67$ and $r^2 = 0.57$ respectively). The results indicated that the groundwater presents a chloride hazard to the soils in the scheme which are becoming saline and therefore require proper management and choice of crops to sustain crop production.

INTRODUCTION

Salinity and sodicity have been extensively reported among the major problems of irrigated agriculture across the world [1-5]. It is estimated that over 900 million hectares of agricultural land are impacted by salinity and sodicity, representing over 6% of all agricultural land and about 20% of the world's irrigated land [6-8]. Soils are classified as saline or sodic based on the soluble salts in saturated extracts (EC) and the proportion of Na to Ca and Mg in saturated extracts (SAR = Na [Ca+Mg/2]½) or the proportion of exchangeable Na to the cation exchange capacity (ESP = Na/CEC/100) [4]. Saline soils have an EC of more than 4 $dS{\cdot}m^{-1}$ and either a SAR of less than 13 or an ESP of less than 9% [9-11]. On the other hand, sodic soils have an EC of less than 4 dSm^{-1} and either a SAR of more than 13 or an ESP of more than 9% [6,12,13].

The methods that are commonly used as indices of salinity and/or sodicity in soils include electrical conductivity, osmotic potential,

Sodium Adsorption Ratio [SAR) and Exchangeable Sodium Percentage (ESP) [3,9,12,14]. SAR measures the proportion of sodium (Na^+) to calcium (Ca^{2+}) and magnesium (Mg^{2+}) and can be used for assessing quality of the water used for irrigation and the status of the soils that receive the irrigation water [1,6]. The effects of high SAR are usually poor soil tilth and soils become sticky when wet resulting in reduced water infiltration [1,15]. The soil tensile strength, a physical measure of the ability of the soils to endure applied forces without being disrupted. is associated with SAR among other factors [13]. Soil pH and physical conditions are also used to determine soil and irrigation water quality in addition to SAR. SAR is commonly used because determining the required CEC for ESP demands laborious and time consuming laboratory tests [12]. The thresholds for SAR and ESP in Zimbabwe are 13% and 9% respectively [16].

The main cause of salinization and sodification is the use of poor quality irrigation water and continued use of this water leads inexorably to increasing salinization and sodification problems and ultimately results in increased cost of production and crop failures [14,17-20]. Salinization is especially serious where groundwater is used for irrigation with the problem being coupled by conversion from natural deep rooted trees and shrubs to commercial shallow-rooted crops that cause the water table to rise and bring more salts with it to the upper layers [8,21]. In addition to effects of crop production, salinity and sodicity have been linked to environmental degradation including loss of below and above ground carbon stocks as vegetation health deteriorate and making soils more susceptible to erosion [9,13,22]. Salinity conditions in agricultural soils have been found to vary over space and seasons [23].

Irrigation water quality and quantity have direct and indirect impact on soil characteristics (physical, biological and chemical) especially in arid and semi-arid regions that unfortunately depend on irrigation [13]. Due to unreliable rainfalls and the climatic gradient, there are over 180 community and government operated surface and sprinkler-based smallholder irrigation schemes in Zimbabwe covering over 10,000 ha producing maize, tomatoes, soya beans, beans, wheat, leafy vegetables and other crops [24-27]. The majority of these are in dry areas and produce food for own use and for marketing and were a result of deliberate policy and institutional frameworks for supporting and promoting irrigated agriculture in dry areas of the country [28].

Water supply to support these schemes is drawn from dams, rivers and/or from deep boreholes with potential for water harvesting to supplement these sources [29-31].

The quality and quantities of the water used on each scheme are as variable as the environmental and geological characteristics in each of the schemes. Water quality and quantity also vary between different seasons and between years as affected by different factors [1]. The variation in water quantity is directly related to productivity in irrigation schemes in the short term while water quality affects productivity in the long term [14,32]. In Zimbabwe, although many schemes have been successful in improving socioeconomic conditions of people in water-limited areas, when assessed in terms of water delivery, agricultural production and socioeconomic indicators, some schemes have not been successful [25,33-35]. The reasons for poor performance or failure in these schemes are both biophysical and socioeconomic [28,32]. This is despite the fact that irrigation schemes are a source of livelihoods for many communal areas in Zimbabwe, the majority of which are in drier areas (mean annual rainfall <650 mm/year) making rainfed crop production risky [21]. Therefore, irrigation schemes are strategic in ensuring food security and agricultural economic activities in areas where crop production is constrained by inadequate or unreliable rainfall.

The biophysical characteristics in schemes need to be optimal for obtaining potential socioeconomic benefits without compromising resource base. This is important for many irrigation schemes, whose demise spells disaster for many people who have little or no alternative options to meet their economic and food security requirements. The objectives of the study were therefore to [1] assess the chemical properties of the irrigation water used at Mutema Irrigation Scheme and [2] determine the levels of selected chemical properties of soils in irrigated and non-irrigated parts of the scheme. It is hoped that knowledge generated from this study will guide agricultural and environmental policy for sustainable irrigation schemes in semi-arid regions.

MATERIALS AND METHODS

Study Area

The study was carried out in Mutema Irrigation Scheme, located in the Chipinge district within the Save catchment. It is 150 km south of Mutare (Figure 1). Altitude is 435 m on a gentle slope of 0.5% from east to west. The scheme was commissioned in 1932 as a pilot irrigation project for communal areas in the area [36,37]. After realising that water abstracted from the Tanganda river was not reliable, four boreholes were drilled in the early 1970s to supplement water supply especially during the winter season [36]. After Cyclone Eline in 2000, the point of abstraction from the Tanganda River was seriously damaged and the boreholes were extended to cover the whole scheme [37]. The soils range from sandy clay loam to sandy loams derived from colluvium and alluvium deposits [36,38]. The scheme is bounded to the south by the Tanganda River, to the west by Save River and to the north by a mountain range. It is in Agro-ecological Region V which is characterized by semi-arid conditions [38]. Mean annual temperatures are above 25°C with hottest months being September and October. Annual total rainfall is very low (less than 600 mm) and is received between December and March [37].

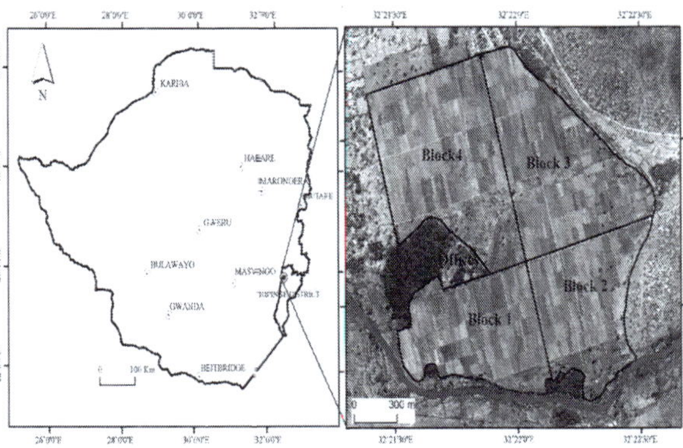

Figure 1: Location of Mutema irrigation schemes in Manicaland and blocks used for sampling. The blocks image was obtained from Google Earth®image.

Soil Sampling

The scheme was divided into four blocks for purposes of soil sampling (Figure 1). From each of the four blocks four soil samples were collected, each from a subdivision of the block into four equal parts. To collect one sample from each block subdivision, four samples were randomly collected and thoroughly mixed to make one composite sample (Figure 1). Therefore, from each block, four samples were collected. Further 4 samples were collected just outside each of the four blocks to be analyzed as the control sample that would indicate the condition of the soils under rainfed system (>100 m from irrigated plots). In total 20 soil samples were collected; 16 inside the irrigated blocks (four from each block) and 4 outside the irrigation scheme. A soil auger was used to collect the samples up to 30 cm depth and all sampling and sample handling were as described by Okalebo et al. [39]. The soil samples were sent for analysis at the Soils and Chemistry Research Institute in Harare. The samples were analysed for pH ($CaCl_2$), EC, K, Ca^{2+}, Mg^{2+}, Na^+ and Total Exchangeable Bases (TEB). The soil samples were collected and analyzed in June 2006 and again in June 2012 using the same procedure.

Water Sampling

Two water samples were collected for analysing the irrigation water to determine its quality; one from the borehole water and another from the adjacent Tanganda River which is a surface source of irrigation water. The water samples were collected on the 12th of July 2012 to coincide with peak periods of irrigation in the scheme. For the borehole sample, two litres of water were collected from the main irrigation line from the borehole in the fields to represent the water that is actually used for irrigation. The water samples were sent for analysis at the Soils and Chemistry Research Institute in Harare. The water samples were analysed for pH, EC, Ca^{2+}, Mg^{2+}, Na^+, HCO_3^- and Cl^-. The quality of the water and the soil were assessed using SAR and ESP, calculated from ionic concentrations of Na^+, Ca^{2+}, Mg^{2+} and CEC.

Data Analysis

Student's t-test was used to analyse for differences in chemical concentrations inside and outside the scheme as well as between chemical concentrations of samples between 2006 and 2012. Correlation analysis (two-tailed) was performed among the variables using the Pearson correlation coefficient while linear regression analysis was conducted to measure the strength of relationships between some variables. All data analysis was done in R environment v. 3.0.1 [40] and all significance tests were at 95% confidence level.

RESULTS

Hydrochemistry of the Irrigation Water

The borehole irrigation water had higher chemical concentrations than the river water for all major cations having double as much Mg^{2+}, six times as much Na^{+} and double the amount of carbonates (Table 1). The borehole water had five times as much Cl^{-} than the river water, and there was a higher conductivity (1729 Msm/cm) in borehole water compared to river water (594 Msm/cm) and this indicated a potential salinity hazard. Although the SAR value was more than four times as much in borehole water (2.16) than in river irrigation water (0.47), there were not much difference in pH values for the borehole and river water which were slightly acidic to neutral (6.8 and 6.7 respectively).

Table 1: Chemical properties of water samples from the river and borehole irrigation water

Attribute	River	Borehole
Ca2+ (mg/1)	2.4	3.35
Mg2+ (mg/1)	2.25	4.75

Na+ (mg/1)	0.72	4.35
HCO_3^- (mg/1)	1.95	3.95
Cl- (mg/1)	0.78	3.6
Conductivity (Msm/cm)	594	1729
pH	6.7	6.8
SAR	0.47	2.16
Ca:Mg	1.07	0.71

Chemical Characterization of Soils

Soil pH and Electrical Conductivity

The general soil conditions in Mutema Irrigation Scheme were alkaline with a high pH ranging from 8 to 8.2 for the irrigated parts of the scheme and 7.2 outside the scheme in 2006 (Table 2). For the irrigated blocks, the highest mean pH was for Block 1 (8.2) while the lowest was for Block 3 which had a mean pH of 8.0. In 2012, mean pH values were slightly lower than in 2006 but still reflecting alkaline soils with mean values for the irrigated parts of the scheme above 7.8 while the mean for samples collected outside the scheme was 7.6 (Table 2). For the irrigated blocks the highest mean pH was for Block 3 (7.9) while the other 3 blocks had a mean pH of 7.8. There were significant differences in soil pH between soils in the scheme and those outside the scheme for 2006 ($t = 5.48$, $p < 0.05$) and 2012 ($t = 6.0$, $p < 0.05$).

There were no significant differences in EC between the irrigated and non-irrigated blocks in 2006 ($t = 0.87$, $p > 0.05$, $df = 18$). In 2006, the highest EC for irrigated blocks was recorded in Block 2 (222.5) while the lowest was for Block 1 (145), which was even lower than that of non-irrigated rainfed parts (153.8). However, samples collected in 2012 indicated that the EC of the blocks inside the scheme is higher than that from the non-irrigated parts of the scheme ($t = 5.25$, $p < 0.05$). Results for 2012 indicated that Block 3 and 4 had the highest EC values (380 and 360 respectively) while the lowest for the irrigated blocks was in Block 1 (250), with samples collected outside the scheme having significantly lower EC in 2012 (Table 2).

Soil SAR and ESP

There were highly significant differences in SAR between the irrigated blocks and the non-irrigated soils in the scheme for samples collected in 2006 ($t = 3.98$, $p < 0.001$, $df = 18$). Block 1, 2 and 3 showed high SAR values that were far above the maximum threshold of 13% in 2006 (Table 2). In 2012, Block 4 had very low SAR values (2.89%) which were lower than outside the scheme (11.08%) explaining the lack of significant differences between the irrigated and the outer parts of the scheme ($t = 0.62$, $p > 0.05$). In terms of ESP of the topsoil, only Block 3 surpassed the threshold of 9% in 2006, indicating that the soils were sodic in 2006. Like for SAR, all irrigated blocks had higher ESP than the control samples in 2006 but Block 4 had the lowest in 2012 (Table 2).

Exchangeable Cations and Ratios

There were no significant differences in K, Mg^{2+}, Ca^{2+} and TEB between the irrigated and non-irrigated soils ($p > 0.05$) in the scheme (Tables 3 and 4). Results for 2006 indicated that Block 1 had the lowest of all exchangeable cations while for 2012; it was Block 4 which had the lower concentrations. Results were not consistent about changes in concentrations of cations between 2006 and 2012. The soil samples collected outside the scheme had significantly lower Na^{+} content than those collected from inside the irrigation scheme ($p < 0.05$, $t = 3.20$, $df = 18$). Except for Block 1 and Block 5 (outside), Na^{+} concentrations increased in the scheme between 2006 and 2012 (Table 4).

Table 2: Mean values of soil pH, EC, SAR and ESP for Mutema Irrigation Scheme for 2006 and 2012

Block Irrigation		pH		EC (Msm/cm)		SAR		ESP (%)	
2006		2012	2006	2012	2006	2012	2006	2012	
1	Irrigated	8.2	7.8	145.0	250	15.57	21.04	4.7	3.9
2	Irrigated	8.1	7.8	222.5	300	25.17	15.48	5.9	3.3
3	Irrigated	8.0	7.9	165.0	380	36.85	20.69	9.1	3.6
4	Irrigated	8.1	7.8	195.0	360	32.14	2.89	7.4	0.6
5	Rainfed	7.2	7.6	153.8	90	8.49	11.08	1.2	2.1

Table 3: Mean concentrations of soil cations K^+ and Mg^{2+} in Mutema irrigation scheme and their change between 2006 and in 2012

Block	Irrigation		K+ (mg/kg)			Mg2+ (mg/kg)	
		2006	2012	Chan ge	2006	2012 % Change	
1	Irrigated	0.3	0.75	158.6	2.8	7.55	167.7
2	Irrigated	0.6	0.77	36.3	4.2	5.76	37.7
3	Irrigated	0.8	1.4	71.8	4.9	7.73	56.9

4	Irrigated	1.2	0.85	—30.8	4.3	4.33	1.9
5	Rainfed	1.9	1.28	—31.6	3.9	3.58	-8.7

Table 4: Mean concentrations of soil Ca^{2+} and Na^{+} in Mutema irrigation scheme and their change between 2006 and in 2012

Block			Ca2+ (mg/kg)		Na+ (mg/kg)		
	Irrigation	2006	2012	Change	2006	2012	Change
1	Irrigated	7.5	20.27	170.3	0.5	1.11	122.0
2	Irrigated	13.6	14.12	3.9	1.1	0.69	—35.8
3	Irrigated	13.0	23.16	78.2	1.7	1.15	—32.7
4	Irrigated	15.0	15.91	5.9	1.5	0.13	-91.4
5	Rainfed	24.1	21.98	-8.8	0.3	0.56	61.2

There was an increase in Na^{+} with increase Mg^{2+} in ($R^2 = 0.65$, $p < 0.001$, Figure 2) and Ca^{2+}($R^2 = 0.57$, $p < 0.001$, Figure 3) in the soi s collected inside the scheme. The soil samples showed that the ratio of Ca^{2+} to Mg^{2+} was lower in Block 1 and Block 3 (2.6) while highest in samples outside the scheme (6.1) (Table 4).

DISCUSSION

The analysis of groundwater used for irrigation in Mutema Irrigation Scheme has indicated high levels of chloride, bicarbonates and sodium, presenting a hazard to the soils and crops in the scheme. The so ls in the schemes had higher pH and exchangeable Na^+, ESP and SAR compared to samples collected outside the scheme suggesting that the groundwater was the main source of contamination for the soils in Mutema Irrigation Scheme. Using irrigation water with a bicarbonate hazard results in the pH of the soil rising continually and this will ultimately result in chloroses in plants and other crop production problems [16, 19].

The results indicate that approximately 14.4 kg of Cl^- are added to the soils in Mutema Irrigation Scheme per hectare per cropping cycle (assuming a cropping cycle of 4 months for vegetables and grains supplied at recommended rates of 25 mm per week [21]. Therefore between 2006 and 2012, about 172 kg $Cl^- \cdot ha^{-1}$ have been added to the soils of the scheme from the borehole irrigation water (assuming a double cropping cycle per year). This rate of salt build up will largely depend on soil drainage in the scheme. Higher Cl^- deposition in agricultural areas has been reported in other areas [5]. According to Mapanda and Mavengahama [21], the maximum permissible soil chloride for crops such as tomatoes and legumes which are common in the scheme is 10 mg/l.

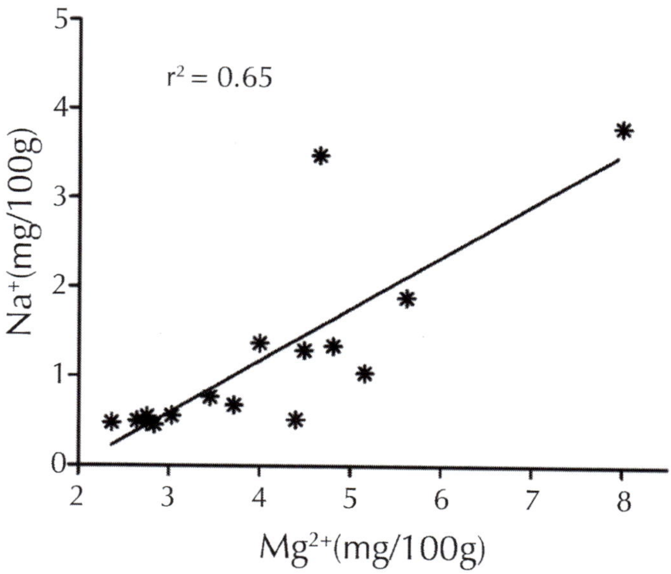

Figure 2: Relationship between Mg^{2+} and Na^+ in the soils of Mutema Irrigation Scheme.

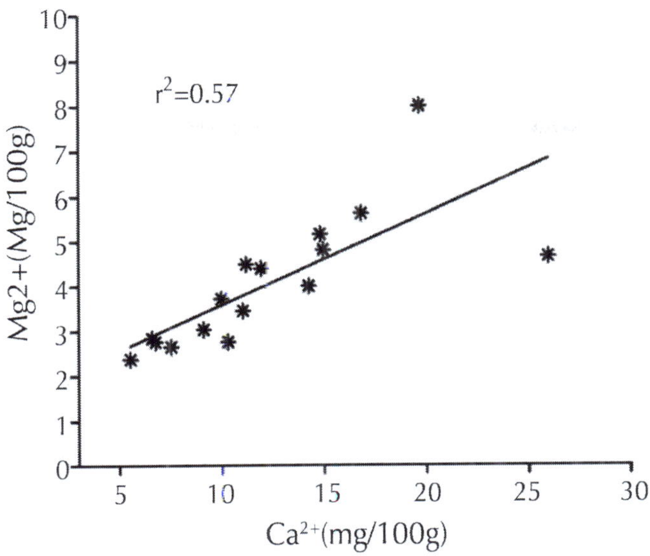

Figure 3: Relationship between Mg^{2+} and Ca^{2+} in the soils of Mutema Irrigation Scheme.

The higher concentration of chloride in the groundwater has potential to affect production of crops given the cumulative nature of chloride in the soils. For example, Ünlükara et al. [11] determined threshold values of 1.5 and 6.7 $dS \cdot m^{-1}$ in irrigation water for fruit yield and vegetative dry weight of eggplant (Solanum melongena L.) in field experiments where it was observed that increasing soil salinity results in reduced water uptake and resultantly reduce growth and yields. Similar findings were reported in other crops such as tomatoes, okra, maize and other crops produced in Mutema Irrigation Scheme [41-45]. Unlike the river water that flows, the higher concentration in the borehole water is attributed to the Save aquifer system that obtains the high soluble salts and sodium from the underlying geological formations [16]. The effects of water and soil quality on crop productivity therefore need to be further investigated in the scheme.

The results indicate that the soils in the scheme are alkaline to the extent that may not support sustainable crop production in some blocks. The soils in the area generally have high pH levels (above 7 for outside the scheme) but irrigation using saline water exacerbates the situation. This is indicated by the fact that the conditions inside the scheme have higher pH, EC, SAR and ESP as shown in by the 2006 and

2012 results (except for Block 4 in 2012). High salinity levels in the scheme as measured by these attributes may be affecting crop plants physically and chemically. Physically, high salinity levels change the osmotic potential between the soils and roots of the plants, thereby reducing the capacity of the plants to absorb water from the soils and transport it through its system for development and functioning of essential parts of the plant system [1, 9, 17, and 46]. Chemically, high concentrations of some specific ions such as Na^+, HCO_3^-, and C^-can be toxic to some crops such as vegetables, maize and wheat, which are sensitive to salinity [1,6]. A soil sample taken in 1990 showed that the soils were already becoming alkaline with pH ranging from 7.2 to 7.8 [36].

In 2012 the results indicated lower concentrations in Block 4, which were lower than outside the scheme. This could be attributed to the lack of cropping in the block between 2006 and 2012 as much of the block was left fallow due to inadequate irrigation since Block 4 is at the end of the line. This means that there is a possibility of improving the soils in the scheme through proper management as shown by the changes in Block 4 between 2006 and 2012. Given the higher SAR percentages in the scheme that exceed the threshold of 13%, it is important that in the short term use corrective treatments are applied in some blocks that are becoming sodic as part of the cropping system for farmers in the scheme. These could include crop rotation, organic soil amendments and use of gypsum and other ameliorants [8,47]. However, these conditions have implications on soil microbial activity and decomposition while increasing erosion hazard and therefore the option of using organic soil amendments maybe slower than expected [9,10].

The effects of the high SAR percentages are that the soil hydraulic system is affected, as aggregates will begin to break down resulting in poor soil structure. This will make the soils less productive as they will be sticky when wet and crusty when dry making tillage operations very difficult [14,15]. In addition, the ion ratios have shown that Mg^{2+} exceeds Na^+ and this indicates that there is excess Na^+ which is known to result in poor drainage [46]. Ca^{2+} is known to counter the dispersion effects of Na^+ thereby counteracting the adverse effects of excessive Na^+ in the soils [6,10,14]. It is not only the high amounts of soluble salts that have adverse effects on the soil and crops but also high concentrations of specific ions including sodium (1). When both are not favourable for crop production, there will be ripple effects on crop productivity.

The problems of salinity in the scheme could result in poor water and nutrient uptake by a plant which is usually interpreted as lack of adequate irrigation or fertilization (1). This is because the saline conditions in the scheme reduce the osmotic ability of the crops which in turn reduces the ability of the crops to absorb and transport water even when available, resulting in physiological droughts [1,44]. Thus farmers would interpret this as lack of sufficient water supply. There is need for development and implementation of agronomic and water management strategies in the scheme to ensure sustainable crop production. The use of surface water to complement the groundwater source should be considered where it can be used as a dilution or a pre-planting flushing of salts to beyond the root zone [5,20,48,49]. Thus, integration of the surface and groundwater sources could help reduce the problem of salinity in the scheme. Given the limitations in water availability and costs associated with pumping from groundwater sources, there is need to consider irrigation methods that minimise the risk of salinity in schemes like Mutema Irrigation Scheme [48,50,51]. A proper water allocation mechanism may be required in order to ensure that all blocks get sufficient water to meet elevated crop water requirements under saline conditions [5,52].

The application of models to simulate the irrigation requirements and crop production potential under the conditions in the scheme is also required to inform crop production and policy on management of existing schemes and development of new ones in dry areas [18, 27,49]. It may also be important to analyze the temporal variations in both water and soil characteristics in the scheme as reported in other studies [1,23]. A broader and regional water quality assessment protocol such as the one suggested by Chilundo et al. [53] may be required to ensure that irrigation schemes remain productive, especially in countries such as Zimbabwe where access to irrigation infrastructure is an important determinant of successful crop production and associated food security and socioeconomic development [34].

CONCLUSIONS

The results have indicated that the groundwater presents chloride and bicarbonate hazards to the scheme. The considerably higher soil pH, EC, SAR and ESP in irrigated blocks directly links the use of groundwater for irrigation to the problems of salinity in the scheme. It is therefore recommended that extension workers should train the farmers on sustainable management practices such as a pre-planting irrigation, use of salt tolerant crop varieties, use of acidifying agents such as Ammonium sulphate and use of crop residues as an agronomic input. There is need for more work in this and other irrigation schemes to determine the link between the soil chemical properties and crop yields.

ACKNOWLEDGEMENTS

We are grateful to AGRITEX extension officers in Mutema irrigation scheme for assisting in collection of soil and water samples in the scheme.

REFERENCES

1. M. S. Yidana, P. A. Sakyi and G. Stamp, "Analysis of the Suitability of Surface Water for Irrigation Purposes: The Southwest and Coastal River Systems in Ghana," Journal of Water Resources and Protection, Vol. 3, No. 1, 2011, pp. 695-710.http://dx.doi.org/10.4236/jwarp.2011.310080
2. A. K. Fasedja, "Eastern and Southern African Agricultural Systems: A Problem Perspecitive," Kilwood Publications, Nairobi, 1999.
3. G Fipps, "Irrigation Water Quality Standards and Salinity Management Strategies," Texas Cooperaive Extension Services, 2007.
4. F. Bouksila, A. Bahri, R. Berndtsson, M. Persson, J. Rozema and S. E. A. T. M. Van der Zee, "Assessment of soil salinization risks under irrigation with brackish water in semiarid Tunisia," Environmental and Experimental Botany, Vol. 92, 2013, pp. 176-185.http://dx.doi.org/10.1016/j.envexpbot.2012.06.002

5. P. Verwey and P. Vermeulen, "Influence of Irrigation on the Level, Salinity and Flow of Groundwater at Vaalharts Irrigation Scheme," Water SA, Vol. 37, No. 2, 2011, pp. 155-164. http://dx.doi.org/10.4314/wsa.v37i2.65861
6. R. Munns, "Genes and Salt Tolerance: Bringing Them Together," New Phytologist, Vol. 167, No. 3, 2005, pp. 645-663. http://dx.doi.org/10.1111/j.1469-8137.2005.01487.x
7. FAO, "Global Network on Integrated Soil Management for Sustainable Use of Salt-Affected Soils," FAO Land and Plant Nutrition Management Service, Rome, 2005.
8. V. N. L. Wong, R. S. B. Greene, R. C. Dalal and B. W. Murphy, "Soil Carbon Dynamics in Saline and Sodic Soils: A Review," Soil Use and Management, Vol. 26, No. 1, 2010, pp. 2-11. http://dx.doi.org/10.1111/j.1475-2743.2009.00251.x
9. R. Setia, P. Gottschalk, P. Smith, P. Marschner, J. Baldock, D. Setia and J. Smith, "Soil Salinity Decreases Global Soil Organic Carbon Stocks (In Press)," Science of The Total Environment, 2012.
10. D. N. Rietz and R. J. Haynes, "Effects of Irrigation-Induced Salinity and Sodicity on Soil Microbial Activity," Soil Biology and Biochemistry, Vol. 35, No. 6, 2003, pp. 845-854.http://dx.doi.org/10.1016/S0038-0717(03)00125-1
11. A. Ünlükara, A. Kurunç, G. D. Kesmez, E. Yurtseven and D. L. Suarez, "Effects of Salinity on Eggplant (Solanum melongena L.) Growth and Evapotranspiration," Irrigation and Drainage, Vol. 59, No. 2, 2010, pp. 203-214.
12. M. Seilsepour, M. Rashidi and B. Khabbaz, "Prediction of Soil Exchangeable Sodium Percentage Based on Soil Sodium Adsorption Ratio," American-Eurasian Journal of Agriculture & Environmental Science, Vol. 5, No. 1, 2009, pp. 1-4.
13. H. Rahimi, E. Pazira and F. Tajik, "Effect of Soil Organic Matter, Electrical Conductivity and Sodium Adsorption Ratio on Tensile Strength of Aggregates," Soil & Tillage Research, Vol. 54, No. 3-4, 2000, pp. 145-153. http://dx.doi.org/10.1016/S0167-1987(00)00086-6
14. R. S. Ayers and D. W. Westcot, "Water Quality for Agriculture," FAO Irrigation and Drainage Paper No. 29, Rome, 1985.
15. A. O. Affullo, "Irrigation Suitability Assessment of Effluents from West Kano Rice Irrigation Scheme, Kisumu, Kenya," Ethiopian

Journal of Environmental Studies and Management, Vol. 2, No. 2, 2009, pp. 1-11.

16. D. Nyamapfene, "The Soils of Zimbabwe," Nehanda Publishers, Harare, 1984.

17. C. Cocozza, C. Pulvento, A. Lavini, M. Riccardi, R. d'Andria and R. Tognetti, "Effects of Increasing Salinity Stress and Decreasing Water Availability on Ecophysiological Traits of Quinoa (Chenopodium quinoa Willd.) Grown in a Mediterranean-Type Agroecosystem," Journal of Agronomy and Crop Science, Vol. 199, No. 4, 2013, pp. 229-240.http://dx.doi.org/10.1111/jac.12012

18. J. Letey and G. L. Feng, "Dynamic versus Steady-State Approaches to Evaluate Irrigation Management of Saline Waters," Agricultural Water Management, Vol. 91, No. 1-3, 2007, pp. 1-10. http://dx.doi.org/10.1016/j.agwat.2007.02.014

19. U. Seeboonruang, "Relationship between Groundwater Properties and Soil Salinity at the Lower Nam Kam River Basin in Thailand," Environmental Earth Sciences, Vol. 69, No. 6, 2013, pp. 1803-1812. http://dx.doi.org/10.1007/s12665-012-2012-5

20. N. Matthews, B. Grové, J. Barnard and L. van Rensburg, "Modelling the Economic Tradeoffs between Allocating Water for Crop Production or Leaching for Salinity Management," Water SA, Vol. 36, No. 1, 2010, pp. 37-44.http://dx.doi.org/10.4314/wsa.v36i1.50905

21. F. Mapanda and S. Mavengahama, "Assessment of Selected Soil Nutrients and Irrigation Water Quality in the Dryland Area of Chivi District, Zimbabwe," Scientific Research & Essays, Vol. 6, No. 14, 2011, pp. 2918-2927.

22. E. Bresler, "Transport of Salts in Soils and Subsoils," Agricultural Water Management, Vol. 4, No. 1-3, 1981, pp. 35-62. http://dx.doi.org/10.1016/0378-3774(81)90043-3

23. C. R. Aza-Gnandji, Y. Xu, L. Raitt and J. Levy, "Salinity of Irrigation Water in the Philippi Farming Area of the Cape Flats, Cape Town, South Africa," Water SA, Vol. 39, No. 2, 2013, pp. 199-210.

24. GOZ, "Support to NEPAD-CADAP Implementation: Bankable Investment Project Profile," Smallholder Irrigation Development, 2004

25. J. M. Makhadho, "An Analysis of Water Management Performance in Smallholder Irrigation Schemes in Zimbabwe," University of Zimbabwe, Harare, 1994.

26. G. Diemer, "Management Processes on Two SmallHolder Sprinkler Layouts: A Case Study from Zimbabwe," Irrigation and Drainage Systems, Vol. 14, No. 3, 2000, pp. 183-197.http://dx.doi.org/10.1023/A:1026525329369

27. I. Samakande, A. Senzanje and E. Manzungu, "Sustainable Water Management in Smallholder Irrigation Schemes: Understanding the Impact of Field Water Management on Maize Productivity on Two Irrigation Schemes in Zimbabwe," Physics and Chemistry of the Earth, Vol.29, No. 15-18, 2004, pp. 1075-1081.http://dx.doi.org/10.1016/j.pce.2004.08.009

28. I. Nyagumbo and J. Rurinda, "An Appraisal of Policies and Institutional Frameworks Impacting on Smallholder Agricultural Water Management in Zimbabwe," Physics and Chemistry of the Earth, Vol. 47-48, 2012, pp. 21-32.http://dx.doi.org/10.1016/j.pce.2011.07.001

29. T. H. Podmore, R. J. McConnen and A. Hungwe, "SmallScale and Smallholder Irrigation in Zimbabwe: Analysis of Opportunities for Improvements," Water Management Synthesis Report 48, 1986.

30. M. M. Sikwela, "Determinants of Household Food Security in the Semi-Arid Areas of Zimbabwe: A Case Study of Irrigated and Non-Irrigated Farmers in Lupane and Hwange Districts," University of South Africa, 2008.

31. G. Nyamadzawo, M. Wuta, J. Nyamangara and D. Gumbo, "Opportunities for Optimization of In-Field Water Harvesting to Cope with Changing Climate in SemiArid Smallholder Farming Areas of Zimbabwe," Springer Plus, Vol. 2, No. 100, 2013, pp. 1-9.

32. N. Mujere, B. Chazovachii, G. Chifodya and A. Mushuku, "Evaluating the Factors Influencing the Variation of Irrigated Wheat Yields: A Case Study of Chinyamutumwa Irrigation Scheme in Zimbabwe," Journal of Sustainable Development in Africa, Vol. 13, No. 14, 2011, pp. 177- 188.

33. E. Manzungu, "Water for Agriculture in Zimbabwe: Policy and Management Options for the Smallholder Sector," University of Zimbabwe, Harare, 1999.

34. H. Bjornlund, "Is Water and Land Redistribution a Driver of Economic Growth and Poverty Reduction? Lessons from Zimbabwe," Water International, Vol. 34, No. 2, 2009, pp. 217-229. http://dx.doi.org/10.1080/02508060902855599

35. K. Nhundu, C. Gwata and A. Mushunje, "Impacts of Zimbabwe European Union Micro-Project Programme (ZIM/EU MPP) in Funding Smallholder Irrigation Projects on Food Security and Income Levels: A Case Study of Mopane Irrigation Scheme in Zvishavane, Midlands Province, Zimbabwe," African Journal of Agricultural Research, Vol. 5, No. 14, 2010, pp. 1759-1771.

36. R. O'Duk and B. Banda, "Irrigation Design for Mutema Irrigation Scheme," FAO and AREX, Harare, 1994.

37. L. Chidoko, "A History of Mutema Irrigation Scheme, 1928-1972," University of Zimbabwe, Harare, 2010.

38. I. P. Anderson, P. J. Brinne, M. Moyo and B. Nyamwanza, "Physical Resource Inventory of Communal Lands of Zimbabwe—An Overview," NRI Bulletin 60, NRI, Harare, 1993.

39. vM. T. Dastorani, M. Heshmati and M. A. Sadeghzadeh, "Evaluation of the Efficiency of Surface and Subsurface Irrigation in Dryland Environments," Irrigation and Drainage, Vol. 59, No. 2, 2010, pp. 129-137.

40. D. Love, P. van der Zaag, S. Uhlenbrook and R. J. S. Owen, "A Water Balance Modelling Approach to Optimising the Use of Water Resources in Ephemeral Sand Rivers," River Research and Applications, Vol. 27, No. 7, 2011, pp. 908-925.

41. S. van der Kooij, M. Zwarteveen, H. Boesveld and M. Kuper, "The Efficiency of Drip Irrigation Unpacked," Agricultural Water Management, Vol. 123, 2013, pp. 103-110.http://dx.doi.org/10.1016/j.agwat.2013.03.014

42. W. de Lange, B. Mahumani, M. Steyn and S. Oelofse, "Monetary Valuation of Salinity Impacts and Microbial Pollution in the Olifants Water Management Area, South Africa," Water SA, Vol. 38, No. 2, 2012, pp. 241-248. http://dx.doi.org/10.4314/wsa.v38i2.9

43. H. C. Komakech, M. Condon and P. van der Zaag, "The Role of Statutory and Local Rules in Allocating Water between Largeand Small-Scale Irrigators in an African River Catchment," Water SA, Vol. 38, No. 1, 2012, pp. 115-126.http://dx.doi.org/10.4314/wsa.v38i1.14

44. M. Chilundo, P. Kelderman and J. H. Ókeeffe, "Design of a Water Quality Monitoring Network for the Limpopo River Basin in Mozambique," Physics and Chemistry of the Earth, Vol. 33, No. 8-13, 2008, pp. 655-665.

Chapter 4

Trace Element Concentration and Speciation in Selected Mining-Contaminated Soils and Water in Willow Creek Floodplain, Colorado

R. Burt[1], T. Weber[2], S. Park[3], S. Yochum[4], and R. Ferguson

[1]United States Department of Agriculture, Natural Resources Conservation Service (USDA-NRCS), National Soil Survey Center, Lincoln, NE 68508, USA

[2]USDA-NRCS, Gainesville, FL 32606-6677, USA

[3]USDA-NRCS, Denver, CO 80225-0426, USA

[4]USDA-NRCS, Fort Collins, CO 80526, USA

ABSTRACT

Long-term mining activities in the mountains around Creede, Colorado have resulted in significant contamination in soils and water in the Willow Creek floodplain. Total major and trace were determined for soils and water and sequential chemical extraction for soils. Objectives were to determine concentrations and potential reactivity of trace elements and investigate their relationship with other soil and water properties. Water trace elements showed significant variability among sites, ranging from 347 to 12108 μg/L. Relative trend showed (Zn > Sr > Ba) > (Mn > W > Cd) > (Sn > V ≈ Ni ≈ Cu > Co) > (Ag). Soil trace elements showed significant short-range spatial variability, ranging from 2819 to 19274 mg/kg. Relative trend showed (Pb ≈ Zn > Mn > Ba > P) > (As > Cu > Sr > V > Cd > Sb ≈ Ag) > (Co ≈ Cr > Mo ≈ Sn ≈ Ni) > (Be ≈ W > Se ≈ Hg). Predominant fractions were oxide, specifically-sorbed/carbonate bound, and residual. Water soluble and exchangeable fractions showed (Zn ≈ Cd) > Pb and Cd > Zn > Pb, respectively. Mobility factors for highly contaminated soils showed Cd ≈ Zn > Pb > Cu > As.

INTRODUCTION

The measurement of the total extractable pool of trace elements has been commonly used to assess the environmental levels or background amounts of trace elements in soils and water [1–7]. Selective fractionations or "geochemical partitioning" have been used to evaluate the potential reactivity of these elements [8–11] and are widely used in soil pollution studies, providing qualitative evidence about trace element reactivity and indirectly of their bioavailability [12–15]. Trace element bioavailability in soils is a complex phenomenon, affected by many factors such as total concentration, pH, organic matter, clay, and redox conditions [7, 16–20]. Trace element toxicity in waters is strongly affected by site-specific water quality factors such as pH, hardness, and other dissolved constituents [21, 22]. An assessment of trace element fate, bioavailability, and transport (e.g., surface and groundwater) is required in order to predict potential contamination and impact upon soil and water quality.

Historic mining activities in the mountains around Creede, Colorado, began around 1889 and continued until 1985. Underground mining of silver and base metals has resulted in Zn and Cd contamination of ground and surface water in and along the broad floodplain of Willow Creek below Creede [23]. Willow Creek, a tributary to the Rio Grande, is polluted from drainage from various mine adits and rock piles upstream of Creede and by leachates from a gravel-capped tailings pile below [23]. Ores common to this area contain Cd which occurs mainly in the Zn sulfides sphalerite and wurtzite and is recovered with Zn usually from polymetallic ores containing Pb and Cu [24].

Major and trace element concentrations (Al, Cd, Cu, Fe, Mn, Pb, and Zn) in river systems are of interest because of their many natural and mine-related sources and their impact on aquatic life through a variety of mechanisms, including both acute and chronic toxic effects of aqueous metals as well as from metal exposure in soils and sediments [22, 23, 25–37].

To assess, monitor, and revitalize the Willow Creek watershed, the Willow Creek Reclamation Committee (WCRC) partnered with numerous state and Federal agencies, including the US Environmental Protection Agency (USEPA); US Forest Service (USFS); US Geological Service (USGS); Colorado Department of Public Health and Environment (CDPHE); Colorado Division of Minerals and Geology; the US Department of Agriculture, Natural Resources Conservation Service (USDA-NRCS), and others. Assessment reports produced by WCRC and/or its partners include [22, 38–43].

The USDA-NRCS designed a sampling scheme to help assess contamination of Willow Creek from nonpoint sources of metals from historical mining activities. In August 2005, USDA-NRCS collected and analyzed soil and water samples from Willow Creek. Initial intent of sampling was to provide background survey levels to help guide potential future samplings for more intensive investigations and later expanded to provide comment on potential soil metal/metalloid bioavailability. This paper reports the results of this investigation. The objectives of this study were as follows: (1) to determine the concentrations and potential reactivity of trace elements using total analysis and sequential chemical fractionation so as to (a) determine the extent and variability of trace element dispersal along the creek and (b) investigate the relationship between trace element concentrations and their potential reactivity with other soil and water properties as well as vegetation or lack thereof along the stream. Knowledge and understanding of these relationships are important to researchers when evaluating trace element data for use in predictive models for environmental purposes. In addition to the geochemistry, detailed site descriptions and other basic soil and water characterization were considered necessary to aid in the interpretation and application of these elemental data.

MATERIALS AND METHODS

Study Area

The Willow Creek watershed is an area of 103 km^2 in Mineral County, Colorado in the northern San Juan Mountains. Willow Creek is formed by the confluence of the East and West Willow Creeks and is a tributary of the Rio Grande River. The City of Creede (Figure 1) and most of the Creede Mining District are located within the watershed. The Creede Mining District is in the central part of the San Juan Volcanic Field. Principal metals produced in the Creede District included Ag, Pb, Zn, Cu, and Au. Production of these metals was from veins along faults that formed during subsidence of the Creede caldera.

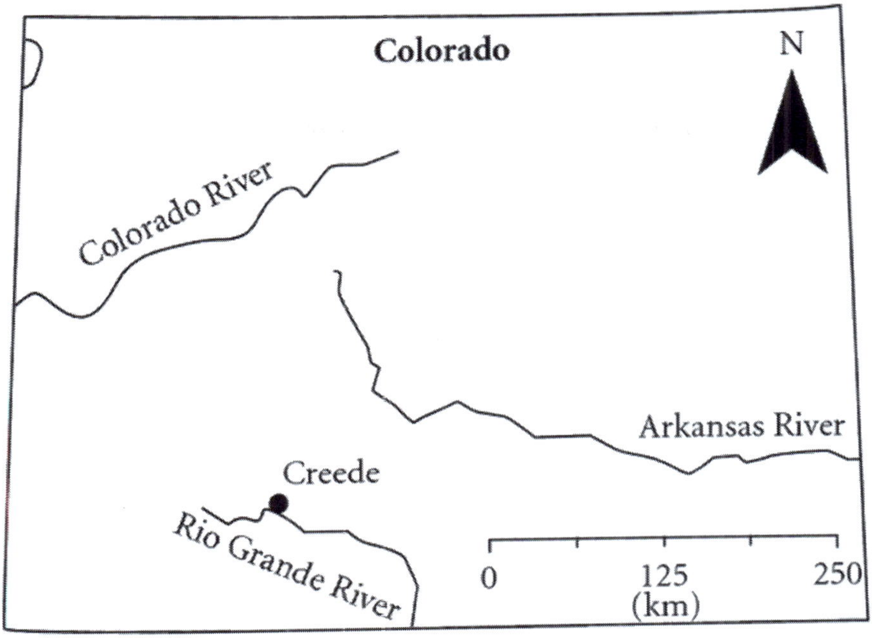

Figure 1: City of Creede, Colorado, part of the historic Creede Mining District, is located within the Willow Creek watershed. Willow Creek is formed by the confluence of the East and West Willow Creeks and is a tributary of the Rio Grande River.

Bedrock geology is dominated by igneous rocks related to extensive tertiary volcanism, with ashflow tuffs the common rock in the watershed above Creede and the dominant rock for mineral deposits in the mining areas along the East and West Willow Creeks [41]. The area beneath Creede and southward to the confluence of Willow Creek and the Rio Grande is predominately the Creede Formation which is mostly lake and river deposits of reworked ash, sand and gravel, and travertine deposits from numerous mineral hot springs forming after the eruption of the Creede Caldera [41]. Much of the older rocks in the lower part of the watershed, especially along the waterways, are covered by Quaternary alluvium from Willow Creek and the Rio Grande [44].

The Willow Creek watershed has a 1.6 km vertical relief and is characterized by steep slopes and distinct vegetative communities. Due to the substantial elevation variability, the climate within the watershed is diverse, affecting the vegetative communities, stream flow magnitude and timing, water temperature, ground-water recharge, and other watershed characteristics [41]. The Willow Creek watershed crosses four major life zones from the high mountains to the Rio Grande Valley, with the origin of Willow Creek in the Alpine Zone (tufted hairgrass-sedge and willow), in the sub-Alpine Zone (Engleman spruce, sub-Alpine fir, Thurber fescue grassland), in Montane Zone (aspen, Douglas Fir, and bristlecone pine), and in the Foothills Zone (Arizona fescue grasslands) [41]. In Creede, the annual average precipitation, temperature, and snowfall are 335 mm, 16°C, and 122 cm, respectively, with the highest and lowest precipitation in August and December and the highest and lowest temperature in July and January, respectively [41]. The largest volume of snow or rain is at the highest elevations. Aquatic resources within the watershed are primarily streams but also include wetlands and riparian areas.

Sampling Strategy: Soils and Water

Soils are naturally variable, their properties changing horizontally across the landscape as well as vertically down the soil profile. Soil map units derived from changes in topography, underlying geology, and dominant vegetation type can be used for horizontal subdivisions, with soil horizons as subdivisions of vertical change [45]. Environmental and/or anthropogenic disturbances introduce additional variation into natural landscapes. In the field, the US soil survey has routinely used pedon sampling to obtain representative samples [46, 47]. In environmental studies as with soil survey, intuitive sampling often forms the basis of exploratory sampling which may be used for qualitative assessment of soils where impact or damage is visible or anticipated [45, 47,48]. Pedon sampling has been used in studies when the area affected is known and/or usually visible, and the types of contaminants are either known or unknown [11, 49–51]. Replicate sampling is often included in sampling plans for water quality, the primary purpose of which is to identify and/or quantify the variability in all or part of the sampling and analysis system [52]. Amount and composition of water samples vary strongly with small changes in location along water body [53]. The primary objective of water sampling is the same as with soil sampling, that is, to obtain a representative sample. For the reasons stated above, the approach to the USDA-NRCS exploratory soil/water sampling at Willow Creek (Aug 9–11, 2005) utilized pedon sampling in the collection of soil samples and water samples replicated on different day (sequential) and same day (concurrent) for laboratory characterization.

All water and soil samples were collected from sites located downstream of Creede on the Willow Creek floodplain. Water sampling was performed in a unidirectional manner with the water flow. Willow Creek has been substantially diverged from its presettlement form due to mining activities [43]. At the time of sampling, Willow Creek was in a braided form. Water sampling was conducted during low-flow conditions when the metal-loading pattern not only shows the metal sources entering the streams on a continual basis but also those sources contributing to high concentrations during the low-flow winter months when mine drainage is less diluted by other water sources and the most toxic conditions likely occur [22, 35, 38]. It was observed that flow conditions in some areas of Willow Creek were free flowing, whereas in other areas, flow was physically obstructed causing water to pool and stagnate, suggesting stream waters had both oxidizing and reducing conditions at the time of sampling.

Water samples were collected from seven sites over a three-day period. Two sites (Westfork and Eastfork) were collected at higher elevations than other sites and are referred to as "baseline" or "reference" samples. Some water samples were collected in conjunction with the pedon sampling [47] and as such bear the same name as the soil. Table 1 provides the names for the water samples and their soil survey identification numbers, with embedded codes indicating concurrent and/or sequential sampling. For example Creek water shows the soil survey numbers S05CO-079-003A-1 through S05CO-079-003C-3, with the A through C representative of sequential samples and 1 through 3 representative of concurrent samples.

Table 1: Selected chemical properties of study waters

Water site	Soil survey no.	EC (dS/m)	pH	Ca	Mg	Na	K	CO3	HCO3	F	Cl	PO4	Br	OAc	SO4	NO2	NO3	
				mmol(+)/L				mmol(−)/L										
Gravelbar water	S05CO-079-001A	0.16	6.6	0.9	0.1	—a	trb	—	0.4	—		tr	—	—	—	1.0	—	—
Gravelbar water	S05CO-079-001B	0.17	6.6	0.8	0.1	—	tr	—	0.4	—		tr	—	—	—	1.0	—	tr
Gravelbar water	S05CO-079-001C	0.16	6.9	0.8	0.1	—	tr	—	0.4	—		tr	—	—	—	1.0	—	—
Cottonwood water	S05CO-079-002A	0.22	6.6	0.7	0.1	—	0.1	—	0.2	tr		0.1	—	—	—	1.6	—	tr
Cottonwood water	S05CO-079-002B	0.22	6.5	0.7	0.1	—	0.1	—	0.2	tr		0.1	—	—	—	1.7	—	tr
Cottonwood water	S05CO-079-002C	0.20	6.6	0.7	0.1	—	0.1	—	0.2	tr		0.1	—	—	—	1.4	—	tr
Creek water	S05CO-079-003A-1	0.14	7.1	0.7	0.1	—	tr	—	0.4	—		tr	—	—	—	0.9	—	—
Creek water	S05CO-079-003A-2	0.14	7.2	0.7	0.1	—	tr	—	0.3	—		tr	—	—	—	0.9	—	—
Creek water	S05CO-079-003A-3	0.15	7.3	0.7	0.1	—	—	—	0.3	—		tr	—	—	—	0.9	—	—
Creek water	S05CO-079-003B-1	0.14	7.3	0.7	0.1	—	—	—	0.3	—		tr	—	—	—	0.9	—	—
Creek water	S05CO-079-003B-2	0.16	6.6	0.7	0.1	—	—	—	0.2	—		tr	—	—	—	0.9	—	—

Creek water	S05CO-079-003B-3	0.15	7.1	0.7	0.1	—	—	—	0.4	—	tr	—	—	—	0.9	—	—
Spruce water	S05CO-079-004A	0.17	7.0	1.0	0.1	—	—	—	0.6	—	tr	—	—	—	0.9	—	—
Spruce water	S05CO-079-004B	0.18	7.1	1.0	0.1	—	—	—	0.6	—	tr	—	—	—	0.9	—	—
Spruce water	S05CO-079-004C	0.18	7.0	1.0	0.1	—	—	—	0.7	—	tr	—	—	—	1.0	—	—
Channel water	S05CO-079-004D-1	0.14	7.3	0.7	0.1	—	—	—	0.3	—	tr	—	—	—	0.9	—	—
Channel water	S05CO-079-004D-2	0.14	7.4	0.7	0.1	—	—	—	0.2	—	tr	—	—	—	0.9	—	—
Channel water	S05CO-079-004D-3	0.14	7.5	0.7	0.1	—	—	—	0.3	—	tr	—	—	—	0.9	—	—
Channel water	S05CO-079-004E-1	0.14	7.4	0.7	0.1	—	—	—	0.3	—	tr	—	—	—	0.9	—	—
Channel water	S05CO-079-004E-2	0.14	7.4	0.7	0.1	—	—	—	0.3	—	tr	—	—	—	0.9	—	—
Channel water	S05CO-079-004E-3	0.14	7.4	0.6	0.1	—	—	—	0.3	—	tr	—	—	—	0.9	—	—
Channel water	S05CO-079-004F-1	0.14	7.4	0.7	0.1	—	—	—	0.4	—	—	—	—	0.7	—	tr	—
Channel water	S05CO -079 -004F-2	0.14	7.4	0.7	0.1	—	—	—	0.3	—	tr	—	—	—	0.9	—	—

Channel water	S05CO -079 -004F-3	0.14	7.4	0.7	0.1	—	—	—	0.4	—	tr	—	—	—	0.9	—	—
Westfork water	S05CO -079 -013	0.11	7.3	0.6	0.1	—	—	—	0.6	—	tr	—	—	—	0.3	—	—
Eastfork water	S05CO -079 -014	0.06	7.5	0.2	tr	—	tr	—	0.4	—	tr	—	—	—	0.1	—	—

[a]— = concentration below detection limits.

[b]tr = trace

Study soils have aquic cond tions, undergoing periods of saturation and reduction [54]. Sampling in August during low-flow conditions provided a good representation of more oxidizing conditions in study soils compared to other times of the year with high-flow rates (high precipitation or snowmelt). Soil surface materials were collected from ten sites, including two sites (Spruce and Railroad), where additional underlying materials were collected. Figure 2 shows sites of these sampled soils. Table 2 provides longitude and latitude for sites and also shows soil name, soil survey identification number, and sampling depth (cm).

Table 2: Soil names, soil survey numbers, depth, and location (latitude and longitude)

Soil	Soil survey no.	Latitude	Longitude
Gravelbar	S05CO -079- 001	37.84070384540	−106.92336102900
Cottonwood	S05CO -079- 002	37.83992953880	−106.92225084200
Spruce	S05CO -079- 004	37.83870504580	−106.92249156600
Railroad	S05CO -079- 005	37.83983640960	−106.92165148000
Western channel	S05CO -079- 015	37.83686982850	−106.92112031400
Salty	S05CO -079- 016	37.83413546890	−106.92016960300
North	S05CO -079- 017	37.83318101160	−106.91817641500

South	S05CO -079-018	37.82851821110	−106.91525304600
Denuded mound	S05CO -079-019	37.82856331670	−106.91564346400
Sedge mound	S05CO -079-020	37.82826082210	−106.91525936500

Figure 2: Willow Creek floodplain showing soil sample sites.

Field and Laboratory Characterization: Soils and Water

Soil and Water Sampling

Soils were described, sampled, and analyzed using standard USDA-NRCS soil survey methods [47, 55]. Water samples were collected using standard USDA-NRCS [47] and USGS [56–58] sampling protocols, expedited for transport to the laboratory, and analyzed using standard USDA-NRCS soil survey laboratory methods [47].

Soil Analyses

Bulk soil samples were air-dried, homogenized, and sieved to <2 mm in the laboratory to remove rock fragments. Samples received standard laboratory characterization performed on air-dried <2 mm soil with resulting data reported on an oven-dry basis. Particle-size analysis was determined by sieve and pipette methods, following pretreatments to remove organic matter (OM) and soluble salts, and chemical dispersion with sodium hexametaphosphate. Total C, N, and S were determined by dry combustion. Soil pH 1 : 1 soil:water was determined. Exchangeable cations (Ca, Mg, K, and Na) were extracted with buffered (pH 7.0) NH_4OAc and measured by atomic absorption spectrometry (AAS). Cation exchange capacity (CEC-7) was determined by Kjeldahl titration of the NH_4-saturated soil. Base saturation by CEC-7 was determined by dividing the sum of NH_4OAc extractable bases by CEC-7 and multiplying by 100. Acidity was extracted by $BaCl_2$-TEA, pH 8.2. Exchangeable Al was extracted with 1 M KCl and determined by inductively couple plasma emission spectrometry (ICP-AES) on samples with pH < 5.5. If electrical conductivity (EC) for 1 : 2 soil water extract was > 0.25 dS/m, a saturated paste extract was prepared and analyzed for cations (Ca, Mg, Na, and K) by AAS and anions (F, Cl, PO_4, Br, OAc, SO_4, NO_2, and NO_3) by ion chromatography and carbonate and bicarbonate (CO_3 and HCO_3) by acid titration. Dithionite-citrate extracts were analyzed for Fe, Al, and Mn (Fe_d, Al_d, and Mn_d, resp.) by AAS. Sodium pyrophosphate extracts were analyzed for Fe, Al, Si, and Mn (Fe_p, Al_p, Si_p, and Mn_p, resp.) by AAS. Acid oxalate extracts were analyzed for Fe, Al, and P (Fe_o Al_o, and P_o, resp.) by ICP-AES. Optical grain counts were conducted on the coarse-silt (0.02–0.05 mm), very fine sand (0.05–0.10 mm), or fine-sand (0.10–0.25 mm) fraction. Crystalline clay minerals (<2 μm) were identified by X-ray diffraction analysis.

Soil Elemental Analyses

Soil samples for major and trace element analysis were prepared metal-free by grinding approximately 10 g of <2 mm soil in a silicon nitride ball mill to <74 μm. Major elements (Al, Ca, Fe, K, Mg, Mn, Na, P, Si, Sr, Ti, and Zr) were determined on soil samples (0.25 g) by microwave digestion (180°C for 9.5 min) combined with 4 mL HF + 9 mL HNO_3 + 3 mL HCl. Following digestion, boric acid was added to neutralize HF (2.5% in 100 mL final volume) and extracts analyzed by ICP-AES. Total extractable (TE) trace elements (Ag, As, Ba, Be, Cd, Cr, Co, Cu, Hg, Mn, Mo, Ni, P, Pb, Sb, Se, Sn, Sr, V, W, and Zn) were determined on soil samples (0.5 g) by microwave digestion (180°C for 7 min) using aqua regia (9 mL HNO_3 + 3 mL HCl). Elements extracted by this method are designated as TE, and the sum of all elements extracted by this method are designated SUM_{TE}. In order to simplify presentation of study results, Mn, P, and Sr are included as trace elements.

Soil Sequential Fractionations

Sequential chemical fractionations were based on the procedure of Tessier et al. [10] and Keller and Vedy [8]. Sequential fractionations were determined on separate samples (0.1 g) in 50 mL centrifuge tubes. Extracts were replicated, with mean values reported. Residues were separated from supernatants by 80-min centrifugation at 4000 rpm. Soils were extracted twice on fractions 3 and 4 with designated reagents (10 mL), and both extracts combined. Washing steps (10 mL H_2O, shake 15 min) were performed between fractionations 2 through 5, with resulting supernatants separated from residues by 80-min centrifugation and solutions combined with initial extracts. Concentrated (16 M) HNO_3 was added to fractions 1 and 2 to acidify final extracts. Trace element concentrations in all extracts were analyzed by inductively coupled plasma mass spectrometry (ICP-MS). Percent recovery was determined by summing the trace elements from the sequential-extraction process (SUM_{SEQ}), divided by the amount determined by the TE elemental analysis by microwave digestion (SUM_{TE}), and multiplied by 100.

Total extractable (TE) and sequential fractions 1 through 6 are designated herein as water soluble (WS), exchangeable (EX), specifically sorbed/carbonate bound (SS/CAR), oxide bound (OX), organic/sulfide bound (OM/S), and residual (RES). The sums of total extractable and sequential fractions are designated herein as SUM_{TE} and SUM_{SEQ}, respectively (Table 3). This designated chemical partitioning of trace elements is based on the evaluations of the sequential fractionation method used herein as applied in various studies [6, 8–12, 59]. In this study, the CH_3COONa/CH_3COOH reagent is used for both calcareous and noncalcareous soils and is designated as SS/CAR. Tessier et al. [10] indicated that the pH 5.0 used in the SS/CAR fraction solubilizes not only the carbonate-bound fraction if present but also a large fraction of specifically adsorbed trace elements from various solid substrates. Similarly, the H_2O_2 extraction (fraction 5) is designated as OM/S, recognizing that not only OM but also sulfide minerals may be extracted to a large extent [10].

Table 3: Designated partition and chemical reagents used for total extractable and sequential extractions[a]

Partition	Reagents/procedure
Total extractable (TE)	9.0 mL 16 M HNO3 + 3.0 mL 12 M HCl (concentrated); microwave digestion: 180 degrees C for 10 min; made to final 50-mL volume with distilled deionized (DDI) water
(1) Water soluble (WS)	10 mL distilled deionized water H2O; shake 1 h
(2) Exchangeable (EX)	10 mL 0.1 M NaNO3; shake 1 h
(3) Specifically sorbed/ carbonate bound (SS/CAR)	10 mL 1 M CH3COONa, adjusted to pH 5 with CH3COOH; shake 5 h
(4) Fe-Mn oxide bound (OX)	20 mL 1 M NH2OH·HCl in 25% v/v CH3COOH; shake 4 h

(5) Organic/ sulfide-bound (OM/S)	2 mL 0.02 M HNO3 + 5 mL 30% H2O2, adjusted to pH 2 with concentrated HNO3; heat to 85 degrees C for 2 h with occasional agitation; add 3 mL H2O2; heat to 85 degrees C for 2 h with occasional agitation; cool and wash with 5 mL 3.2 M NH4OAc in 20% v/v HNO3; dilute to 20 mL; shake 30 min
(6) Residual (RES)	9.0 mL 16 M HNO3 + 3.0 mL 12 M HCl (concentrated); microwave digestion: 180 degrees C

[a]After Tessier et al., 1979 and modified by Keller and Vedy, 1994.

Water Analyses

Water samples were analyzed for cations (Ca, Mg, Na, and K) by AAS and anions (F, Cl, PO_4, Br, OAc, SO_4, NO_2, and NO_3) by ion chromatography and carbonate and bicarbonate (CO_3 and HCO_3) by acid titration. Electrical conductivity and pH were determined on water samples. Water samples were filtered (0.45 μm), acidified with HCl, and analyzed for major (Al, Ca, Fe, K, Mg, and Sr) and trace elements (Ag, As, Ba, Be, Cd, Co, Cr, Cu, Mn, Mo, Ni, P, Pb, Sb, Sn, Sr, V, W, and Zn) by ICP-AES. For purposes of this study, Sr is included as trace element.

Quality Control

Quality control (QC) included blanks, duplicates, certified reference material (CRM), and standards from the National Institute of Standards and Technology (NIST). Reference material for soil samples included CRM Loam Soil C, high-purity standards, Charleston, SC and NIST 2709a San Joaquin Soil. The high purity standard CRM-TMDW was used for water samples. Elemental recoveries and relative standard deviation (RSD) for soil and water QC standards ranged from 90% to 115% and ≤11%, respectively. Percent recovery for QC soil standards, determined by dividing SUM_{SEQ} by SUM_{TE} and multiplying by 100, ranged from 124 to 130 percent.

RESULTS AND DISCUSSION

Water: Chemical Properties

Water samples were less acidic than most of the sampled soils, with pH ranging from 6.5 to 7.5 (slightly acid to slightly alkaline) (Table 3). Of the sampled waters, the Channel and Creek water samples were generally less acidic than those waters sampled in conjunction with the soils (Gravelbar, Cottonwood, and Spruce). The relatively lower pH for the water associated with the sampled soils and for the water from the second day sampling of the Creek water may be due in part to the capturing of more sediment in the sample compared to the other samplings. In general, there are relatively small differences in water pH within sites, that is, same-day samples (concurrent) and different-day samples (sequential). For example, Channel water showed a mean and standard deviation of 7.5 ± 0.05 over a three-day period. These data suggest minor daily fluctuations in the stream or channel pH may be anticipated under relatively similar environmental conditions. This lack of high daily variability within the time frame of sampling may also indicate that the water system has a reasonably good buffering system. Poorly buffered waters and high daily variability in pH are often attributed to biological processes that affect the carbonate buffering system.

The water pH for the two locations at the higher elevations designated as "reference" or "baseline" sites (Westfork and Eastfork) were 7.3 and 7.4, respectively. These data suggest the pH of the water at the upper elevations was not dramatically different from the water at the lower elevations (excluding those associated with sampled soils) during the time frame of sampling. As water moves through soils and rocks, the pH may increase or decrease as additional chemical reactions occur. The pH within streams can impact toxic materials, with high acidity or alkalinity tending to convert insoluble metal sulfides to soluble forms, thereby increasing the concentrations of toxic metals.

Other data [43] showed significant pH variability in groundwater across the Willow Creek floodplain, ranging from 2.6 to 7.0. Yochum et al. [43] suggests pH can be a good indicator for groundwater contamination, that is, the link between low pH and groundwater contamination can be used for early detection for ground water contamination plume movement, minimizing the expense of water quality analyses.

Electrical conductivity (EC) showed variability among sites, with the lowest EC values in the "reference" sites, with 0.06 and 0.11 dS/m in the Eastfork and Westfork waters, respectively (Table 3). Highest EC values were found in those waters sampled in conjunction with the soils, ranging from 0.16 to 0.22 dS/m. Data showed good agreement within sites, with the Channel and Creek waters showing the greatest consistency with means and standard deviations of 0.15 ± 0.008 and 0.14 ± 0.000 dS/m, respectively, over a three-day period. Similar to pH, these data suggest minor fluctuations in the EC of the stream or channels may be anticipated under relatively similar environmental conditions. Additionally, these data may indicate that the salt load may increase in transport by stream waters from higher elevations to the valley floor and that these salts would be more significant in areas where water stagnates or pools with soils and sediments.

Predominant ions in water samples were $SO_4 > Ca > HCO_3$ (Table 3). Ion concentrations showed good agreement across all samples (among and within sites), with means and standard deviations of 1.0 ± 0.23, 0.8 ± 0.11, and 0.4 ± 0.13 mmol/L for SO_4, Ca, and HCO_3, respectively. Sulfate is a product of pyrite oxidation as well as the oxidation and dissolution of other metal-bearing sulfide minerals. Carbonate buffering results from chemical equilibrium between calcium, carbonate, bicarbonate (HCO_3), carbon dioxide, and hydrogen ions in the water and carbon dioxide in the atmosphere. Buffering causes waters to resist changes in pH [60]. The amount of buffering is primarily determined by carbonate and bicarbonate concentration which are introduced into the water from dissolved calcium carbonate (i.e., limestone) and similar minerals in the watershed [61]. Glucose is the primary product of photosynthesis, with acetate (OAc) forming from the glucose as plants decompose. In comparison, Kimball et al. [38] found the stream water at the injection point in West Willow Creek a $CaHCO_3$-$CaSO_4$ type water, reflecting the chemical weathering of bedrock in the watershed [44]. Kimball et al. [38] further determined these ions and pH remained constant until the inflow of the Nelson Tunnel, at which point the chemical character of water changed to a $CaSO_4$ type, and pH was lower. The Nelson Tunnel discharge is located on West Willow Creek about 2.4 km upstream of Creede and is considered responsible for 75% of the total inorganics (heavy metals) contamination in Willow Creek [42]. With the inflow of East willow creek, pH became more basic again, Ca and SO_4 diluted, and alkalinity remained nearly constant [38]. For this study, soil and water samples were collected below Creede on the southern part of the Willow creek floodplain.

Water: Major and Trace Elements

Metals in natural waters undergo continual changes between dissolved, precipitated, and sorbed-to-sediment forms, with the rate of these adsorption, desorption, and precipitation processes depending on pH, redox potential, water chemistry, and composition of bottom and suspended sediments [62]. These forms have traditionally been subdivided into two fractions, "dissolved" and "particulate", according to an operationally defined limit (0.45 µm) and separated by filtration [63]. Using this partitioning, study waters were filtered with a 0.45-µm filter, with the filtrate analyzed for major and trace elements. In all samples, Fe and Al concentrations were below detection limits (Table 4). The pattern for Mn concentrations was not similar to any other trace element and showed spatial variability, ranging from below detection limits to 173.2 µg/L (Table 5). Other water pollution studies [37, 38, 63, 64] have examined additional fractions (dissolved, colloidal, and particulate), with the dissolved fraction generally characterized as the ultrafiltered (<10 kD). Using this partitioning, Kimball et al. [38] similarly found Al and Fe mostly near the limits of detection and Mn measurable in streams near Creede. Differences between concentrations of Fe and Al compared to Mn in study waters may due in part to the geochemical behavior of these solutes, that is, Fe and Al are among the most reactive constituents likely precipitating and forming colloidal materials, with Mn less reactive and more likely remaining in the aqueous phase.

Table 4: Total major element concentrations for study waters

Water site	Soil survey no.	Al	Ca	Fe	K	Mg	Na	SUMa
		mg/L						(mg/L)

Gravelbar water	S05CO-079-001A	—b	18.5	—	1.2	1.6	5.9	27.2
Gravelbar water	S05CO-079-001B	—	19.2	—	1.1	1.6	6.4	28.4
Gravelbar water	S05CO-079-001C	—	19.3	—	1.2	1.6	6.2	28.2
Cottonwood water	S05CO-079-002A	—	17.7	—	1.9	1.9	11.4	32.9
Cottonwood water	S05CO-079-002B	—	18.1	—	2.1	2.0	10.9	33.1
Cottonwood water	S05CO-079-002C	—	18.4	—	2.1	1.9	9.5	31.9
Creek water	S05CO-079-003A-1	—	17.7	—	1.0	1.7	6.1	26.5
Creek water	S05CO-079-003A-2	—	17.5	—	1.0	1.7	6.0	26.2
Creek water	S05CO-079-003A-3	—	17.3	—	1.0	1.7	6.5	26.4
Creek water	S05CO-079-003B-1	—	17.6	—	0.9	1.7	6.6	26.9
Creek water	S05CO-079-003B-2	—	17.3	—	1.0	1.7	6.5	26.5
Creek water	S05CO-079-003B-3	—	17.4	—	0.9	1.7	6.4	26.4
Spruce water	S05CO-079-004A	—	22.9	—	1.3	1.9	6.1	32.1
Spruce water	S05CO-079-004B	—	23.7	—	1.2	2.0	6.8	33.7
Spruce water	S05CO-079-004C	—	24.6	—	1.3	2.1	6.7	34.6

Channel water	S05CO-079-004D-1	—	17.7	—	1.0	1.7	6.4	26.8
Channel water	S05CO-079-004D-2	—	17.9	—	1.0	1.7	6.4	26.9
Channel water	S05CO-079-004D-3	—	17.5	—	0.9	1.7	6.5	26.6
Channel water	S05CO-079-004E-1	—	17.6	—	1.0	1.7	6.3	26.5
Channel water	S05CO-079-004E-2	—	17.4	—	0.9	1.7	6.2	26.2
Channel water	S05CO-079-004E-3	—	17.1	—	0.8	1.6	6.4	26.0
Channel water	S05CO-079-004F-1	—	17.5	—	0.9	1.7	6.4	26.4
Channel water	S05CO-079-004F-2	—	17.3	—	0.9	1.7	6.2	26.1
Channel water	S05CO-079-004F-3	—	17.2	—	0.9	1.6	6.3	26.1
Westfork water	S05CO-079-013	—	15.0	—	0.7	1.6	5.0	22.2
Eastfork water	S05CO-079-014	—	7.1	—	0.6	0.6	3.7	12.0

[a]SUM = (Al + Ca + Fe + K + Mg + Na).

[b]— = concentration below detection limits

Table 5: Total trace element concentrations for study waters

Water site	Soil survey no.	Ag	As	Ba	Be	Cd	Co	Cr	Cu	Mn	Ni	P	Pb	Sb	Sn	Sr	V	W	Zn	Total conc. by site
										µg/L									µg/L	(µg/L)
Gravelbar water	08SCO-079-001A	0.38	—[a]	150.2	—	18.7	1.5	—	10.3	—	6.0	—	—	—	9.5	242.7	6.0	45.1	4021.1	4495
Gravelbar water	08SCO-079-001B	0.40	—	207.5	—	19.1	1.5	—	8.7	—	5.8	—	—	—	9.7	247.4	6.0	45.1	4027.9	4579
Gravelbar water	08SCO-079-001C	tr[b]	—	169.2	—	18.1	1.4	—	6.4	—	5.8	—	—	—	9.2	248.6	5.9	44.6	4018.4	4528
Cottonwood water	08SCO-079-002A	0.36	—	414.0	0.01	70.3	2.2	—	5.8	141.5	7.1	—	—	tr	9.0	241.9	6.1	164.5	10405.6	11462
Cottonwood water	08SCO-079-002B	0.41	—	314.1	0.01	77.2	2.3	—	5.5	173.2	6.7	—	—	tr	9.4	242.4	6.0	181.9	11082.3	12108
Cottonwood water	08SCO-079-002C	0.36	—	202.6	0.01	62.1	1.9	—	4.5	101.7	6.2	—	—	tr	9.5	239.7	6.1	138.4	9521.3	10294
Creek water	08SCO-079-003A-1	0.42	—	138.8	—	19.4	1.6	—	5.6	76.3	5.7	—	6.7	—	9.3	221.9	6.2	31.1	2963.1	3486
Creek water	08SCO-079-003A-2	0.42	—	134.8	—	18.9	1.6	—	5.7	73.2	5.4	—	8.6	—	9.5	219.7	6.1	31.3	2961.2	3478
Creek water	08SCO-079-003A-3	0.35	—	136.8	—	19.1	1.6	—	5.6	75.4	5.8	—	9.3	—	9.6	217.8	6.0	30.9	2915.1	3433
Creek water	08SCO-079-003B-1	0.38	—	169.8	—	30.1	1.7	—	4.2	83.4	5.9	—	13.5	—	9.4	220.9	6.2	34.7	3198.1	3768
Creek water	08SCO-079-003B-2	0.43	—	178.3	—	24.2	1.6	—	5.8	81.8	5.3	—	149.6	—	9.4	220.0	6.1	41.7	3718.1	4439
Creek water	08SCO-079-003B-3	0.42	—	170.0	—	19.5	1.5	—	4.2	82.8	5.7	—	9.9	—	9.8	222.5	6.1	24.3	3235.4	3801
Spruce water	08SCO-079-004A	0.40	—	150.2	—	24.8	1.4	—	5.9	—	5.8	—	6.8	—	8.7	220.1	5.9	48.8	4335.8	4874
Spruce water	08SCO-079-004B	0.38	—	252.2	—	23.4	1.4	—	6.1	—	5.8	—	5.5	—	9.4	223.1	6.0	47.9	4256.8	4896
Spruce water	08SCO-079-004C	0.45	—	190.0	—	24.6	1.5	—	4.6	—	5.7	—	—	—	9.5	224.6	5.9	50.8	4359.5	4939
Channel water	08SCO-079-004D-1	tr	—	148.4	—	14.1	1.6	—	4.7	69.6	5.3	—	5.6	—	9.1	221.9	6.0	15.7	1650.5	2151
Channel water	08SCO-079-004D-2	tr	—	155.6	—	12.6	1.5	—	5.0	75.4	5.8	—	2.5	—	9.2	217.4	6.1	13.5	1425.5	1968
Channel water	08SCO-079-004D-3	0.41	—	209.0	—	12.3	1.6	—	5.1	67.3	5.4	—	4.0	—	9.5	219.2	6.0	13.6	1472.7	2026
Channel water	08SCO-079-004E-1	tr	—	152.2	—	17.3	1.6	—	4.9	74.8	5.6	—	7.3	—	9.3	221.9	6.1	26.6	2591.9	3119
Channel water	08SCO-079-004E-2	tr	—	148.5	—	17.3	1.6	—	5.3	78.5	5.4	—	6.9	tr	9.4	219.5	6.0	26.4	2570.5	3096
Channel water	08SCO-079-004E-3	tr	—	202.5	—	16.6	1.6	—	5.0	71.7	5.5	—	8.5	—	9.4	218.0	6.0	28.2	2484.2	3055
Channel water	08SCO-079-004F-1	tr	—	188.5	—	19.3	1.7	—	5.6	85.1	5.5	—	5.4	—	9.2	221.8	6.0	31.3	2949.6	3527
Channel water	08SCO-079-004F-2	0.40	—	166.1	—	19.8	1.6	—	5.6	98.6	5.4	—	9.3	—	8.7	220.7	6.0	32.4	3077.5	3643
Channel water	08SCO-079-004F-3	0.35	—	167.7	—	18.2	1.7	—	5.7	84.7	5.2	—	1.5	—	9.5	219.6	6.0	31.7	2941.5	3481
Westfork water	08SCO-079-013	0.38	—	218.5	0.01	2.6	1.4	—	4.8	—	4.5	—	—	tr	9.5	246.1	6.2	—	142.5	654
Eastfork water	08SCO-079-014	0.43	—	107.6	—	1.0	1.4	—	4.3	—	4.2	tr	—	tr	8.7	112.7	6.6	—	98.1	345

[a]— = concentration below detection limits

[b]tr = trace.

Total concentrations of major elements showed slight variability among sites, ranging from 12 mg/L in Eastfork to 35 mg/L in Spruce (S05CO-079-004B). Overall, Eastfork had lower levels of all major elements compared to other sites. Total major element concentrations were in relatively good agreement across all samples (among and within sites), with mean and standard deviation of 28.3 ± 2.91 mg/L. The relative trend among major elements showed Ca > Na > Mg > K, with good agreement across all samples (among and within sites) showing means and standard deviations of 18.5 ± 2.1, 6.9 ± 1.5, 1.7±0.1, and 1.1 ± 0.4mg/L, respectively.

Total concentrations of trace elements showed significant variability among sites, ranging from 345 to 12108 g/L, with much of this disparity explained by Zn levels (Table 5). Total trace element concentrations within sites were in relatively good agreement during the August sampling. It is expected that these concentrations may vary seasonally or year-round. In a study of the upper Animas River watershed in Colorado, affected by acid drainage, Besser et al. [35] found the annual stream discharge (with high runoff in spring and considerably lower discharge in winter) drives the trace element concentrations in the water column, resulting in aqueous concentrations of dissolved Zn and Cu typically near annual minima during summer and reaching annual maxima in late winter when these concentrations may be greater by a factor of three or more [65].

Water geochemistry showed a wide suite of trace elements present across the sampling area. The relative trend among trace elements showed (Zn > Sr > Ba) > (Mn > W > Cd) > (Sn > V ≈ Ni ≈ Cu > Co) > (Ag). Overall water composition, in terms of elements of potential concern, is in good agreement with the reports by the Willow Creek Reclamation Committee [39, 40].

Zinc concentrations ranged from 98.1 and 142.3 g/L in Eastfork and Westfork waters, respectively, to 11088.3 g/L in Cottonwood, S05CO-079-002B, suggesting significant metal loading downstream to Willow Creek. Zinc had the highest concentrations of all metals in study waters. Kimball et al. [38] similarly found East Willow Creek an insignificant source of metal loads and Zn concentrations the highest of all metals in the Willow creek watershed, with its cumulative in stream load one of the highest of all the watersheds in the rocky mountains [66–71]. Zinc under oxidizing conditions and in the absence of other anions with which to react tends to remains as a dissolved cation, but in the presence of certain reactants, Zn can form sulfates, oxides/hydroxides, carbonates, and phosphates whose solubilities depend more on pH than on redox potential [62]. August sampling of Willow Creek showed SO_4 ranging from below detection limits to 1.7 mmol/L and the absence of Fe and Al. Study water pH ranged from 6.5 to 7.5, with those waters with pH (Gravelbar, S05CO-079-001A-C and Cottonwood, S05CO-079-002A-C) rendering the highest Zn values of 4027.9 and 11088.3 g/kg, respectively. Total concentrations of colloidal Fe and Al in the Animas river near Silverton, Colorado, were considered limiting factors to Zn adsorption during the summer months, with Zn adsorption greater downstream in association with pH increases and during spring when Fe and Al concentrations were higher [37]. Studies that show significant Zn adsorption in the water column over a narrow range of about one pH unit near pH 7 have a large excess of absorbent, usually Fe or Al hydroxides, relative to the total Zn concentration [37, 72, 73].

Copper levels ranged from 3.6 to 10.3 μg/L. Cadmium which most often exists in small quantities in Zn ores ranged from 1.0 to 77.2 μg/L. Lead was detected in some samples, with the greatest amount (149.6 μg/L) in one replicate of the second-day sampling of the Creek water (S05CO-079-003B-2). This higher Pb value may be due in part to the capturing of more sediment in the sample compared to the other samplings and/or reflective of time or length scales of temporal or lateral variability in the stream. In a study of streams near Creede, Kimball et al. [38] found colloidal and dissolved Pb relatively high compared with many streams affected by mine drainage, with colloidal Pb > dissolved Pb and in higher concentrations downstream from Nelson Tunnel versus upstream despite no great variation in pH or major-ion concentrations. With observed pH range of study waters exhibiting mixed solid phases (organics and inorganics) [39, 40], preferential trace metal adsorption sequence in water system may be Pb > Cu > Cd > Zn [74], explaining in part the lack of Pb in some samples of stream waters compared to Zn.

Strontium and Ba levels showed relatively high levels, ranging from 112.7 to 283.1 and 107.6 to 414.0 μg/L, respectively. Strontium sources are linked to nonore minerals with its loading profiles indicating sources upstream [38]. Study waters showed relatively low Ag levels, ranging from trace to 0.35 μg/L. Arsenic, Be, Cr, P, and Sb were not detected in water samples.

Geochemistry of water samples associated with sampled soils showed higher levels of trace elements compared to stream samplings, for example, Cottonwood water (S05CO-079-002A) with Zn, Cd, and Mn levels of 11088.3, 77.2, and 173.2μg/L, respectively. Cottonwood had some of the lowest pH (6.5 to 6.6) and highest EC (0.20 to 0.2 dS/m) values compared to other sites. Under some circumstances, such as increases in salinity, decreases in redox potential (oxygen deficient), or decreases in pH, metals may be desorbed from sediments and solubilize in the water [75]. These circumstances may include when the water flow is physically obstructed causing water to pool and stagnate. The stream system may have a reasonably good buffering system, with low daily variability in pH and EC. The ability to resist abrupt changes in pH tends to prevent or slow metal conversion to soluble forms, which would increase the concentration of potentially toxic trace elements.

The "reference" or "baseline" sites of Westfork and Eastfork waters had relatively low levels of trace elements compared to other sites with Zn (98.1 and 142.3 μg/L, resp.) and Cd (1.0 and 2.6 μg/L, resp.). Overall, Eastfork showed lower levels of trace elements compared to Westfork. These data indicate trace element concentrations are significantly different at these higher elevations compared to the sampling sites at the lower elevations.

Soils: Physical, Mineralogical, and Chemical Properties

Study soils were classified as Aquic Cryofluvents, showing redox depletions with low chroma within the upper 50 cm of surface, and having aquic conditions for some time in normal years [54]. Soil textures were primarily coarse-textured (loam to coarse sand), with the >2-mm fraction ranging from 1 to 73% (Table 6). Soils had mixed mineralogy (kaolinite, smectite, and hydroxyl-interlayer smectite) (Table 6). Smectitic clays typically show the finest particle-size distribution of naturally occurring clay minerals (particles <0.05 μm) and have electrically charged surfaces that permit the adsorption and absorption of water and other substances, whereas the kaolin group of clays includes particles with relatively large size and neutral electrical charge resulting in less potential for persistent suspension and turbidity. Smectitic clays have been linked as a primary cause of persistent turbidity (fine inorganic and organic particles suspended within water column) in the Cascade Range, North Santiam River, Willamette National Forest [76]. For August and July surface and mine water sampling, total suspended solids (TSS) ranged upwards of 15.4, 18.1, and 12.8 mg/L in main channel sites in East Willow Creek, West Willow Creek, and Mainstream Willow Creek, respectively [39].

Table 6: Selected chemical and physical properties of study soils

Soil	Depth	ECa	pH	CEC-7b	Base Sat.c	Total C	Total N	Total S	Bray P-1	Fe_d^d	Al_o^e	Mnd	Sand	Silt	Clay	>2-mm fraction
	(cm)	dS/m		(cmol/kg)	(%)	%			mg/kg	%			%			(%)
Gravelbar	0–73	–f	6.8	6.9	86	0.47	0.060	0.04	22.5	1.1	0.04	0.2	91.9	5.0	3.1	65
Cottonwood	0–90	–	5.4	15.2	47	1.19	0.134	0.04	21.4	1.0	0.07	tr	69.0	18.3	12.7	19
Spruce	0–10	–	6.4	12.9	85	2.05	0.308	0.14	23.9	1.3	0.07	0.1	73.1	17.0	9.9	24
	10–40	–	6.7	6.8	97	0.43	0.708	0.10	30.8	1.3	0.04	0.1	88.7	6.4	4.9	63
Railroad	0–61	–	5.9	7.0	19	0.52	0.035	0.22	13.9	1.6	0.05	0.1	80.3	13.8	5.9	59
	61–120	–	4.9	5.3	42	0.23	0.040	0.13	21.0	1.4	0.03	0.2	89.6	6.8	3.6	66
Western Channel	0–40	–	6.4	6.2	84	0.41	0.139	0.17	23.9	0.9	0.03	0.1	89.5	5.9	4.6	70
Salty	0–20	8.00	5.7	22.4	29	2.76	0.229	0.28	–	1.3	0.09	0.1	51.4	32.5	16.1	1
North	0–40	–	6.2	7.1	100	0.50	0.063	0.19	10.1	1.3	0.07	0.2	87.0	7.8	5.2	73
South	0–40	–	6.1	6.0	57	0.27	0.034	0.08	9.6	1.4	0.03	0.1	93.4	3.6	3.0	47
Denuded Mound	0–40	–	4.6	16.1	34	1.33	0.185	0.23	17.0	1.2	0.08	tr	55.8	28.5	15.7	3
Sedge Mound	0–40	–	5.6	23.1	100	2.43	0.246	0.13	26.3	1.0	0.07	tr	50.8	30.0	19.2	4

EC^a = electrical conductivity, saturated paste extract.

CEC-7^b= cation exchange capacity by NH4OAc, pH 7.

Base Sat.[c]= base saturation by NH4OAc, pH 7.

= dithionite-citrate extractable iron.

= ammonium oxalate extractable aluminum.

–[f] = concentration below detection limits.

For sampled soils, the pH (1 : 1 water) for surface materials ranged from 5.4 to 6.8 (strongly acid to neutral), with the exception of the Denuded Mound soil with very strongly acidic pH of 4.6 (Table 6). The Sedge Mound soil which was in close proximity to the Denuded Mound soil had a pH of 5.6.

Acidic conditions and low fertility may explain in part the disparity of plant vegetation in the Sedge Mound (100% base saturation by CEC-7, pH 5.6) compared to the Denuded Mound (34% base saturation and 4.6 pH). Total C and N were also lower in the Denuded Mound (1.33 and 0.185%, resp.,) compared to the Sedge Mound (2.43 and 0.246%, resp.,) (Table 6). Additionally, the Sedge Mound compared to other soils showed the highest clay content (19.2%) and highest amount of available P (26.3 mg/kg) as measured by Bray P-1 (Table 6).

The Salty soil showed pH 5.7 with a relatively high saturated paste EC (8.00 dS/m) (Table 6), with the water soluble anion sulfate (168.1 mmol(−)/L) in greatest amount (data not shown). Typically, these water soluble cations and anions would show greater ion balance, suggesting some cations present in significant amounts were not measured by this method, with potential candidates of Fe, Cu, and Zn sulfates. Melanterite (hydrated Fe sulfate) is one of only a few wate-soluble sulfate minerals, forming white to green encrustations and crystal aggregates. The primary source of Fe for melanterite is Fe sulfates such as pyrite, marcasite, and chalcopyrite. The chemical processes involved in mining that lead to widespread acidification of water bodies and deposition of heavy metals have been studied extensively, with some researchers [77] finding that microbes coat the pyrite surface with exudates that dramatically speed up oxidation and production of sulfuric acid instead of protecting the surfaces from oxidation as had been previously expected.

Soils: Total Extractable Major and Trace Elements

Total concentrations of major elements showed good agreement among soils. The relative trend among major elements showed Si > Al > K > Fe > Ca > Mg > Na > Ti > Sr > Zr, with good agreement among soils (Table7). This trend was also in relatively good agreement with a dataset of 486 pedons from across the US, inclusive of both anthropogenic and nonanthropogenic soils such as those affected by mining activities, with data showing Si > Al > K > Fe > Ca > Mg > Na ≈ Mg > Zr > Ti [6].

Table 7: Total major element concentrations for study soils

Soil	Depth	Al	Ca	Fe	K	Mg	Na	Si	Ti	Zr	SUMa
	(cm)	g/kg									(g/kg)
Gravelbar	0–73	64	10	26	45	4	15	327	2	98	591
Cotton wood	0–90	74	14	26	33	6	18	314	3	100	588
Spruce	0–10	69	14	31	35	5	16	308	3	94	575
	10–40	62	14	38	37	4	17	324	3	88	587
Railroad	0–61	52	8	35	39	3	13	336	2	86	574
	61–120	55	9	32	41	3	15	339	3	88	585
Western Channel	0–40	63	12	34	38	4	16	336	3	92	598
Salty	0–20	66	12	30	28	6	14	312	3	108	579
North	0–40	66	11	30	40	4	16	332	2	93	594
South	0–40	62	12	35	37	4	17	338	3	84	592
Denuded Mound	0–40	66	13	30	28	6	16	332	3	92	586
Sedge Mound	0–40	72	17	25	29	7	17	306	3	96	572

SUM[a] = (Al + Ca + Fe + K + Mg + Na + Si + Ti + Zr).

Relatively high amounts of K in these coarse-textured soils were related to naturally high levels of potassium feldspars. Optical analysis showed the same trend for predominant minerals among all soils as follows: Chert > Plagioclase Feldspar > Quartz > Potassium Feldspar.

Total Fe and Al ranged from 25 to 38 and 52 to 74 g/kg, respectively. These values exceed the geometric means for Fe and Al (22 ± 2 and 40 ± 2 g/kg, resp.) as determined by Burt [6]. Dithionite-citrate and ammonium oxalate extractions [78–82] showed "active" Fe, Al, and Mn components, with Fe_d 1.0% to 1.6%, Al_o 0.03% to 0.09%, and Mn_d trace to 0.2% (Table 6). The Fe_d and Al_o are generally considered good measures of the total pedogenic Fe and Al in soils, respectively [79]. The Fe_d/Fe_t ratios in soils showed free Fe oxides comprise 38% to 46% of the total Fe pool. Data suggest study soils serve as seasonal sources of Fe- and Al-rich colloids to stream waters. The interface between soil surface horizons and streams becomes tightly linked during high-flow events, resulting in increases in stream concentrations and depletions of soil constituents [83].

Total concentrations of trace elements were significantly greater in soils than in water samples. Relatively low concentrations of metals in the Hei River, China despite the contamination to soils and sediments were related to water flow rates high enough to dilute releases from the stream bank soils and sediments [84]. Total trace element (TE) concentrations in study soils showed significant short-range spatial variability, ranging from 2819 in Sedge Mound to 19274 mg/kg in Salty, and seeming to follow no directional pattern along the stream (Table8). Most of this variability was explained by differences in Zn and Pb levels among soils, varying by factors of nearly 20 and 30, respectively.

Table 8: Total extractable trace elements (TE) of study soils

Sample no.	Depth	Ag	As	Ba	Be	Cd	Co	Cu	Cr	Hg	Mn	Mo mg/kg	Ni	P	Pb	Sb	Se	Sn	Sr	V	W	Zn	SUM^a_{TE} (mg/kg)
Gravelbar	0-73	8.5	98.4	522.8	0.8	15.2	6.5	85.2	4.5	0.10	1805.4	2.5	2.7	454.5	2622.8	7.9	0.37	2.1	39.6	37.6	0.69	2415.0	8133
Cottonwood	0-90	2.0	35.1	289.2	1.2	21.1	5.8	26.6	6.4	0.05	618.0	1.3	3.3	552.8	1154.4	3.2	0.50	1.3	57.8	44.0	0.17	2648.9	5473
Spruce	0-10	11.7	153.7	1138.4	1.3	25.6	7.5	112.7	5.9	0.17	1719.2	4.7	3.3	593.0	3776.6	11.6	0.51	3.3	59.0	45.0	1.03	3316.9	10991
	10-40	10.1	183.7	890.0	0.8	14.7	8.1	93.3	6.6	0.13	2130.0	5.3	3.6	561.5	3251.3	13.2	0.36	3.5	53.0	58.9	1.12	2635.9	9925
Railroad	0-61	30.0	559.6	817.2	0.9	17.0	5.5	125.3	5.5	0.30	1310.6	12.9	2.6	463.7	4977.2	46.2	0.31	5.2	36.8	41.1	2.40	2656.7	11117
	61-120	32.8	368.5	874.8	0.7	8.3	5.5	80.1	5.2	0.28	1528.9	6.6	2.5	476.9	3134.5	31.8	0.31	3.8	40.4	41.9	1.08	1248.4	7893
Western Channel	0-40	10.6	157.0	476.0	0.9	23.9	7.4	98.8	5.2	0.14	2113.0	3.5	2.8	523.1	3702.9	14.0	0.41	8.8	44.4	45.8	1.25	4043.4	11283
Salty	0-20	23.2	168.1	617.4	1.5	86.6	9.7	110.8	8.1	0.27	1675.1	4.3	4.4	695.1	4807.7	14.0	0.63	5.6	92.3	46.1	0.52	10903.0	19274
North	0-40	14.1	172.8	956.9	1.0	24.9	9.0	128.5	4.3	0.20	2786.6	3.3	3.0	448.7	4233.7	15.6	0.40	2.9	49.2	39.2	1.38	3767.0	12663
South	0-40	17.3	167.2	639.6	0.6	13.7	6.6	57.2	5.3	0.18	1324.1	5.0	2.6	509.7	2701.3	13.5	0.37	6.1	50.1	48.4	0.95	2093.9	7664
Denuded Mound	0-40	10.7	82.9	339.3	1.1	25.9	5.4	60.3	5.9	0.16	786.2	3.7	3.2	747.3	8748.5	7.1	0.71	1.6	73.1	40.0	1.46	4371.8	15317
Sedge Mound	0-40	0.9	25.2	343.7	1.4	4.5	6.1	17.2	8.1	0.04	627.4	0.7	4.4	733.0	306.6	1.7	0.69	1.2	100.5	45.8	0.06	589.8	2819

[a] $SUM_{TE} = (Ag_{TE} + As_{TE} + Ba_{TE} + Be_{TE} + Cd_{TE} + Co_{TE} + Cu_{TE} + Cr_{TE} + Hg_{TE} + Mn_{TE} + Mo_{TE} + Ni_{TE} + P_{TE} + Sb_{TE} + Se_{TE} + Sn_{TE} + Sr_{TE} + V_{TE} + W_{TE} + Zn_{TE})$

[a] $SUM_{TE} = (Ag_{TE} + As_{TE} + Ba_{TE} + Be_{TE} + Cd_{TE} + Co_{TE} + Cu_{TE} + Cr_{TE} + Hg_{TE} + Mn_{TE} + Mo_{TE} + Ni_{TE} + P_{TE} + Sb_{TE} + Se_{TE} + Sn_{TE} + Sr_{TE} + V_{TE} + W_{TE} + Zn_{TE})$

Trace elements from anthropogenic sources have been found to be more mobile than those of native origin [85–87]. A common method to quantify the release and transport of metals in surface–contaminated soils is the distribution of these contaminants within the soil profile [88]. Total concentrations were higher in surface layers of study soils compared to underlying materials, with Spruce showing 10991 versus 9925 mg/kg, respectively, and Railroad 11117 versus 7893 mg/kg, respectively. These data are in agreement with Burt [6] finding the median value of most trace elements (Cd, Cu, Ni, Pb, Zn, and Hg) significantly higher in surface horizons of anthropogenic soils (n = 392) compared with those in surface horizons of nonanthropogenic soils.

Similar to the water data, there was a wide suite of trace elements present in the soils across the sampling area. The relative trend among trace elements showed (Pb ≈ Zn > Mn > Ba > P) > (As > Cu > Sr > V > Cd > Sb ≈ Ag) > (Co ≈ Cr > Mo ≈ Sn ≈ Ni) > (Be ≈ W > Se ≈ Hg), with good agreement among soils. Geochemistry of soils does not parallel the geochemistry of the stream waters. Relatively high Pb amounts in the soils compared to the water samples may be reflective of its relatively low solubility in water and low mobility in soils, accumulating primarily on the surface where its increasing presence can begin to affect soil microflora [89].

The Denuded Mound and Sedge Mound soils were demonstrative of the vegetative disparities along the stream bank. The Denuded Mound and Sedge Mound showed significant differences in total trace element concentrations (15317 versus 2819). Soil pH for the Denuded Mound and Sedge Mound was 4.6 versus 5.6, respectively. These results would tentatively suggest that a correlation may exist between metal levels and soil pH and vegetation or lack thereof along the stream. In a study of smelter–contaminated soils (Cu, Zn, Pb, and Cd) in Anaconda and Deer Lodge Valley, Montana, Burt et al. [49] found that low pH, alteration in species composition in plant communities, and reduction in plant growth and vigor often occur in conjunction with soil contamination, but that these easily measurable and observable properties alone could not be used as field indicators of total trace element concentrations. Observable differences in these Montana plant communities were determined to serve more appropriately as criteria to determine impact classes (e.g., severe, moderate, and slight) for soils affected by contamination but not correlated to total trace element content [49]. Soil pH is useful information in terms of bioavailability of trace elements [90–92].

Soils: Chemical Fractionations

Percent recovery ranged from 104 to 140%, indicating relatively good agreement between SUM_{SEQ} and SUM_{TE}. There was also good reproducibility between the two replications of the sequential fractionation (data not shown). Precision and accuracy of the sequential extraction procedure are generally considered inherently good, with the limiting factor the inherent heterogeneity of the specimen [10]. Study soils were located on the floodplain with colluvial and alluvial deposits which create a more heterogeneous medium for soil development (e.g., horizonation, morphology, and texture) [93]. For more details, see Table 9.

Table 9: Sequential extraction as percent of total extractable trace elements (TE) in study soils

Soil			Sequential fractionationsa					Recoveryb
	Depth	WS	EX	SS/CAR	OX	OM/S	RES	
	(cm)	% of total						(%)
Gravelbar	0–73	1	3	42	45	10	11	113
Cottonwood	0–90	1	21	40	36	9	12	119
Spruce	0–10	1	5	47	41	11	21	126
	10–40	1	3	41	68	9	19	140
Railroad	0–61	2	9	16	47	14	24	111
	61–120	1	11	23	48	6	25	113
Western Channel	0–40	1	4	41	42	14	10	112
Salty	0–20	16	16	43	28	5	7	115
North	0–40	1	4	37	47	14	10	113
South	0–40	4	10	31	38	10	15	108
Denuded Mound	0–40	8	22	42	15	9	7	104
Sedge Mound	0–40	2	6	28	54	12	22	124

[a]Sequential fractionations: WS, water soluble; EX, exchangeable; SS/CAR, specifically sorbed/carbonate bound;

OX, oxide–bound; OM/S, organic /sulfide–bound; RES, residual.

[b]Recovery = (SUMSEQ/SUMTE) × 100.

In general, study soils showed the OX (15 to 68%), SS/CAR (23 to 47%), and RES (7 to 25%) as the predominant fractions over the EX (3 to 22%), OM/S (5 to 14%), and WS (1 to 16%) fractions. Element–specific bioavailable fractions (WS and EX) showed Pb (0.29% to 4.60% and 0.94% to 24.27%, resp., see Table 10), Cd (0.99% to 23.37% and 5.28% to 40.18%, resp., see Table 11), Zn (0.77% to 24.39% and 4.14% to 32.67%, respectively, see Table 12), and As (0.09% to 2.15% and 0.39% to 2.73%, resp., see Table 14). In contrast, the Montana smelter soils [11] had lower total metal concentrations (1836 to 3605 mg/kg), smaller WS (0.02% to 1.0%) and EX (0.2% to 9.3%) fractions, larger RES (14.3% to 70.1%) fraction, and smaller potentially bioavailable fractions (Pb, Cd, and Zn), exception being the greater EX Cu fraction in the Montana soils (4.67% to 30%) compared to Willow Creek (1.01% to 5.32%, Table 13). Similarly, the Montana smelter soils showed the OX fraction (29.7% to 39.8%) as the predominant fraction, with this fraction exceeding the RES fraction in surface layers in both studies. These data suggest that some anthropogenic inputs were added, as well as later transformed to, OX forms. Anthropogenic metals, initially present as poorly crystalline phases, may oxidize in surface soil horizons. Oxides have a scavenging action for trace elements disproportionate to their own concentrations, and as such constitute important sources of potentially bioavailable trace elements [10]. Overall, results suggest the Willow Creek soils are more contaminated than the Montana smelter soils, showing not only greater total metal levels but also an overall greater pool of potentially bioavailable trace elements, specifically those of concern (e.g., Cd, Pb, and Zn).

Table 10: Sequential fractionation of Pb in study soils

Soil	Depth (cm)	PbWS	PbEX	PbSS/CAR	PbOX	PbOM/S	PbRES	Pb^{a}_{SUMSEQ} (mg/kg)
		mg/kg						
Gravelbar	0–73	9.2	33.9	1756.2	773.3	13.6	157.8	2744
Cotton wood	0–90	4.1	11.9	903.5	287.0	10.3	58.6	1275
Spruce	0–10	11.9	39.9	2641.5	1194.3	24.2	228.6	4140
	10–40	12.6	34.7	2108.9	981.4	19.4	302.3	3459
Railroad	0–61	60.9	168.9	1097.5	2240.7	40.4	1351.5	4960

	61–120	24.6	175.2	1086.7	1056.9	19.1	865.4	3228
Western Channel	0–40	17.4	123.5	2529.5	1045.8	24.1	176.9	3917
Salty	0–20	22.1	50.2	3603.0	1267.1	16.1	223.9	5182
North	0–40	9.9	83.7	2625.0	1595.8	17.3	221.9	4554
South	0–40	126.9	148.7	1461.8	673.5	20.7	321.7	2753
Denuded Mound	0–40	284.2	2070.0	5628.9	402.5	13.7	130.3	8530
Sedge Mound	0–40	1.0	1.8	164.4	90.9	2.9	27.3	288

$$^{a}Pb_{SUMSEQ} = Pb\,Sum(WS + EX + SS/CAR + OX + OM/S + RES).$$

Table 11: Sequential fractionation of Cd in study pedons

Soil	Depth (cm)	CdWS	CdEX	CdSS/CAR	CdOX	CdOM/S	CdRES	Cd^{a}_{SUMSEQ} (mg/kg)
		mg/kg						
Gravelbar	0–73	0.21	1.4	6.9	1.5	3.8	0.4	14
Cotton wood	0–90	0.24	6.1	10.7	1.3	1.5	0.2	20
Spruce	0–10	0.54	3.5	14.5	1.7	4.7	0.5	25
	10–40	0.15	1.2	6.4	3.2	3.8	0.6	15
Railroad	0–61	0.63	4.7	1.7	1.3	8.2	0.7	17
	61–120	0.26	3.2	1.6	1.3	1.4	0.3	8
Western Channel	0–40	0.25	1.9	8.4	2.4	8.3	1.3	23
Salty	0–20	19.44	21.1	35.9	3.0	2.9	0.9	83
North	0–40	0.41	2.2	8.5	2.8	9.6	0.9	24
South	0–40	0.69	3.7	2.9	1.4	3.7	0.9	13
Denuded Mound	0–40	5.36	8.4	2.7	0.8	7.1	0.6	25
Sedge Mound	0–40	0.08	0.21	2.79	0.27	0.58	0.05	4

$$^{a}Cd_{SUMSEQ} = Cd\,Sum(WS + EX + SS/CAR + OX + OM/S + RES).$$

Table 12: Sequential fractionation of Zn in study soils

oil	Depth (cm)	ZnWS	ZnEX	ZnSS/ CAR	ZnOX	ZnOM /S	ZnRES	Zn^{a}_{SUMSEQ} (mg/kg)
		mg/kg						
Gravelbar	0–73	22.0	113.3	1170.6	568.8	602.5	142.0	2619
Cotton wood	0–90	42.0	949.3	1051.4	496.1	273.7	93.3	2906
Spruce	0–10	66.0	304.8	1696.2	819.8	722.6	1129.4	4739
	10–40	20.3	111.4	1233.3	1169.0	586.7	747.5	2635
Railroad	0–61	89.6	534.4	237.1	551.4	1329.6	255.3	2997
	61–120	34.5	345.0	277.0	370.7	227.4	152.7	1407
Western Channel	0–40	22.9	183.1	1639.0	913.5	1364.0	300.1	4423
Salty	0–20	2959.0	2850.1	3932.7	1646.9	501.0	244.6	12134
North	0–40	32.4	173.0	1373.0	822.0	1515.3	228.6	4144
South	0–40	105.9	390.1	539.9	435.1	556.8	185.1	2213
Denuded Mound	0–40	855.0	1222.6	476.9	674.8	1190.2	225.9	4645
Sedge Mound	0–40	13.2	32.3	255.3	143.4	122.5	87.3	654

$$^{a}Zn_{SUMSEQ} = Zn\,Sum(WS + EX + SS/CAR + OX + OM/S + RES).$$

Table 13: Sequential fractionation of Cu in study soils

Soil	Depth (cm)	CuWS	CuEX	CuSS/ CAR	CuOX	CuOM /S	CuRES	Cu^{a}_{SUMSEQ} (mg/kg)
		mg/kg						
Gravelbar	0–73	0.6	1.3	20.9	16.9	41.5	10.5	92
Cottonwood	0–90	0.6	0.4	3.6	9.1	7.7	9.1	31
Spruce	0–10	1.2	1.5	26.5	35.3	47.1	15.3	127
	10–40	0.7	1.1	22.1	30.7	44.9	11.3	111
Railroad	0–61	1.8	3.5	17.1	62.8	19.4	28.7	133
	61–120	0.8	4.4	15.9	31.4	10.9	19.6	83
Western Channel	0–40	0.6	2.9	27.5	22.6	46.6	8.6	109
Salty	0–20	1.0	1.3	38.2	38.0	24.0	15.7	118

North	0–40	0.4	2.4	31.9	45.3	40.6	15.0	135
South	0–40	2.5	2.5	12.1	14.5	15.1	9.4	56
Denuded Mound	0–40	0.6	1.0	12.1	20.3	13.0	11.1	58
Sedge Mound	0–40	0.4	0.6	0.5	4.0	4.2	8.6	18

[a] $Cu_{SUMSEQ} = Cu\,Sum(WS + EX + SS/CAR + OX + OM/S + RES)$.

Table 14: Sequential fractionation of as in study soils

Soil	Depth (cm)	AsWS	AsEX	AsSS/CAR	AsOX	AsOM/S	AsRES	As^{a}_{SUMSEQ} (mg/kg)
		mg/kg						
Gravelbar	0–73	0.6	1.8	15.4	70.5	4.6	53.2	146
Cotton wood	0–90	0.2	0.3	3.5	34.0	2.3	15.6	56
Spruce	0–10	0.7	2.2	25.7	143.7	8.7	70.7	252
	10–40	0.9	2.7	25.0	214.5	8.0	103.9	355
Railroad	0–61	5.7	7.6	18.6	520.5	10.3	294.5	857
	61–120	2.2	10.4	24.7	187.5	5.5	151.2	381
Western Channel	0–40	0.8	4.2	23.6	125.8	15.6	77.1	247
Salty	0–20	0.3	1.6	23.9	180.2	3.8	80.0	290
North	0–40	0.5	3.4	25.2	169.9	8.8	78.5	286
South	0–40	4.5	4.2	21.8	83.4	2.3	95.1	211
Denuded Mound	0–40	0.1	0.5	10.2	73.4	1.5	34.2	120
Sedge Mound	0–40	0.3	0.5	2.3	21.9	1.7	10.6	37

[a] $As_{SUMSEQ} = As\,Sum(WS + EX + SS/CAR + OX + OM/S + RES)$.

Soils in both studies have been affected by over 100 years of anthropogenic inputs, with the Willow Creek soils from mining and the Montana study soils from smelting of ores, primarily copper. While both sites have been inactive for a number of years, the geochemical processes are more active and impacts on–going in Willow Creek compared to the Montana study soils, related to site–specific differences, for example, topography, geology, climate, soil properties, and contaminant type. The Willow Creek soils are located on a floodplain whereas the Montana soils are on uplands with steep slopes (e.g., 30%), suggesting more active processes of metal transport to and potential accumulation in the Willow Creek soils. Smelter–derived particles have been shown to rapidly alter soon after deposition through weathering processes, leading to a release of metals in soils [94]. The aquic conditions of the soils along Willow Creek provide for a more dynamic system of metal release and further transport into the soil–water environment compared to the aridic soils in Montana. The higher pH Montana soils (derived from calcareous alluvium) provide for a greater buffering against acidification, resulting in a less soluble and therefore less mobile trace element fraction compared to the strongly acid to neutral pH Willow Creek soils (derived from ashflow tuffs). Soil pH and redox conditions are considered important factors affecting the chemistry of metals in soils and their uptake by organisms, with alternating aerobic and anaerobic conditions leading to changes in both factors thereby affecting most of those processes regulating metal speciation in soils [95–99].

All study soils showed high levels of contamination, with the Salty and Denuded Mound soils being the most contaminated showing the greatest SUM_{TE} (19274 and 15317 mg/kg, resp.) and some of the greatest WS (16 and 8%, resp.) and EX (16 and 22%, resp.) fractions. The Salty soil with pH 5.7 and EC 8.00 dS/m had the highest Cd levels of total, WS, and EX fractions, with 84, 19.4, and 21.1 mg/kg, respectively. In general, soluble Cd levels are greater in moderately acid, oxidizing conditions, increasing 10 fold under these conditions compared to reducing conditions at all pH levels [75]. Cadmium has also been shown to be affected by salinity, with increased salinity supporting increased Cd concentration in solution [75]. The Salty soil also had the highest levels of Zn with 12134, 2959.0, and 2850.1 mg/kg in the total, WS, and EX fractions, respectively.

Metal mobility was evaluated using the mobility factor (MF) [59, 100], representing the proportion of metals in the water soluble and exchangeable fractions and calculated as follows:

$$MF_x = \left(\frac{(WS_x + EX_x)}{WS_x + EX_x + OX_x + (SS/CAR_x) + RES_x} \right) \times 100, \qquad (1)$$

Where x = element.

The MF was assessed for the highly contaminated salty and denuded soils as follows: Cd ≈ Zn > Pb > Cu > As. Values ranged as follows: 48.7 to 55.3% Cd; 44.7 to 47.9% Zn; 1.4 to 27.6% Pb; 1.9 to 2.8% Cu; and 0.5 to 0.6% as. These MFs are in relatively good agreement with other studies [101, 102], with Cd and Zn showing the highest mobility. Low CEC, sandy nature, and high permeability of study soils indicate inherently low sorption capacity for Cd and Zn, and as such high mobility. Generally, in all study soils, Cd appears mainly in the SS/CAR form, whereas Zn in the SS/CAR, OX, and OM/S forms. The relatively high MF_{Pb} in the Denuded soil may be due in part to having the lowest OX fraction (15%) compared to other study soils, and similarly a low OM fraction (9%). Both of these fractions (Fe–Mn oxide and organic) have been shown to adsorb or complex Pb in less available forms in soils [103–105]. Generally, in all study soil, Pb appears mainly in the SS/CAR and OX forms. The low MF_{As} for the salty and denuded soils may be due in part to its greater association with the OX and RES fractions over the more soluble fractions. These results show good agreement with other studies investigating As forms and distribution in soils [106–109].

CONCLUSIONS

Historic mining activities and impacts from these activities have resulted in significant fluvial contamination of sampled soils and water in the Willow Creek floodplain. Thorough understanding of the factors affecting this contamination requires a good knowledge of the local geology, geomorphology, hydrology, climate, soil properties, and contaminant amount, type, and form. There is a wide suite of trace elements present in both soils and waters. Trace element forms in water are strongly affected by site-specific water quality factors such as pH, EC, and other dissolved constituents. Total concentrations of trace elements in water showed significant variability among sites, with much of this disparity explained by Zn levels. Soil properties such as pH and redox conditions affect chemical forms and potential reactivity. Total trace element concentrations in study soils showed significant short-range spatial variability, and seemed to follow no directional pattern along the stream. In general, study soils showed the OX, SS/CAR, and RES as the predominant fractions over the EX, OM/S, and WS. The relative distribution of trace elements within the more soluble fraction (e.g., WS and EX) of soils is similar to their respective total concentrations, suggesting a relationship between total trace elements and the concentration of soluble, mobile fraction. Largest element-specific bioavailable fractions (WS and EX) were found in Cd, Zn, and Pb.

The expected outcome of the Willow Creek study and similar studies of anthropogenic soils conducted by USDA Soil Survey (e.g., [6, 7, 11, 49–51, 110]) is to better characterize important soils in specified–project areas, and more generally establish important relationships with other soil properties as well as with other soils both uncontaminated and contaminated. Our knowledge of the various soil processes affecting the elemental amounts and distribution in US soils and their relationships with other soil properties will enhance the understanding of the fate and transport of anthropogenic elements, thereby expanding the utility and application of soil survey knowledge in areas of environmental concern (e.g., urban, mine spoil reclamation, munitions disposal, smelter emissions, and agricultural waste applications).

REFERENCES

1. H. T. Shacklette and J. G. Boerngen, "Element concentrations in soils and other surficial materials of the conterminous United States," U.S. Geological Survey Professional Paper 1270, U.S. Government Printing Office, Washington, DC, USA, 1984.
2. L. M. Shuman, "Fractionation method for soil microelements," Soil Science, vol. 140, no. 1, pp. 11–22, 1985.
3. G. G. S. Holmgren, M. W. Meyer, R. L. Chaney, and R. B. Daniels, "Cadmium, lead, zinc, copper, and nickel in agricultural soils of the United States of America," Journal of Environmental Quality, vol. 22, no. 2, pp. 335–348, 1993.
4. A. R. Mermut, J. C. Jain, L. Song, R. Kerrich, L. Kozak, and S. Jana, "Trace element concentrations of selected soils and fertilizers in Saskatchewan, Canada," Journal of Environmental Quality, vol. 25, no. 4, pp. 845–853, 1996.
5. M. Chen, L. Q. Ma, and W. G. Harris, "Baseline concentrations of 15 trace elements in Florida surface soils," Journal of Environmental Quality, vol. 28, no. 4, pp. 1173–1181, 1999.
6. R. Burt, M. A. Wilson, M. D. Mays, and C. W. Lee, "Major and trace elements of selected pedons in the USA," Journal of Environmental Quality, vol. 32, no. 6, pp. 2109–2121, 2003.
7. M. A. Wilson, R. Burt, S. J. Indorante et al., "Geochemistry in the modern soil survey program,"Environmental Monitoring and Assessment, vol. 139, pp. 151–171, 2008.

8. C. Keller and J. C. Vedy, "Distribution of copper and cadmium fractions in two forest soils," Journal of Environmental Quality, vol. 23, no. 5, pp. 987–999, 1994.
9. G. Rauret, R. Rubio, J. F. Lopez-Sanchez, and E. Casassas, "Determination and speciation for metal solid speciation in heavily polluted river sediments," International Journal of Environmental Analytical Chemistry, vol. 35, pp. 89–100, 1988.
10. A. Tessier, P. G. C. Campbell, and M. BIsson, "Sequential extraction procedure for the speciation of particulate trace metals," Analytical Chemistry, vol. 51, no. 7, pp. 844–851, 1979.
11. R. Burt, M. A. Wilson, T. J. Keck, B. D. Dougherty, D. E. Strom, and J. A. Lindahl, "Trace element speciation in selected smelter-contaminated soils in Anaconda and Deer Lodge Valley, Montana, USA,"Advances in Environmental Research, vol. 8, no. 1, pp. 51–67, 2003.
12. L. Q. Ma and G. N. Rao, "Chemical fractionation of cadmium, copper, nickel, and zinc in contaminated soils," Journal of Environmental Quality, vol. 26, no. 1, pp. 259–264, 1997.
13. A. Chlopecka, J. R. Bacon, M. J. Wilson, and J. Kay, "Forms of cadmium, lead, and zinc in contaminated soils from southwest Poland," Journal of Environmental Quality, vol. 25, no. 1, pp. 69–79, 1996.
14. R. M. Harrison, "Chemical association of lead, Cd, Cu, and Zin in street dusts and roadside soils,"Environmental Science and Technology, vol. 15, pp. 1378–1383, 1981.
15. S. S. Iyengar, D. G. Martens, and W. P. Miller, "Distribution and plant availability of soil zinc fractions,"Soil Science Society of America Journal, vol. 45, pp. 735–739, 1981.
16. B. R. Singh, "Soil pollution and contamination," in Advances in Soil Science: Methods for Assessment of Soil Degradation, R. Lal, W. H. Blum, C. Valentin, and B. A. Stewart, Eds., pp. 279–299, CRC Press, Boca Raton, Fla, USA, 1997.
17. S. M. Reichman, "The response of plants to metal toxicity: a review focusing on copper, manganese, and zinc," Australian Mineral and Energy Environment Foundation, Occasional Paper No. 14, 2002.

18. G. Renella, P. Adamo, M. R. Bianco, L. Landi, P. Violante, and P. Nannipieri, "Availability and speciation of cadmium added to a calcareous soil under various managements," European Journal of Soil Science, vol. 55, no. 1, pp. 123–133, 2004.
19. S. Suave and D. R. Parker, "Chemical speciation of trace elements in soil solution," in Chemical Processes in Soils, M. A. Tabatabai and D. L. Sparks, Eds., Soil Science Society of America Book Series No. 8, pp. 655–688, Madison, Wis, USA, 2005.
20. J. J. D'Amore, S. R. Al-Abed, K. G. Scheckel, and J. A. Ryan, "Methods for speciation of metals in soils. A review," Journal of Environmental Quality, vol. 34, no. 5, pp. 1707–1745, 2005.
21. J. M. Diamond, E. L. Winchester, D. G. Mackler, W. J. Rasnake, J. K. Fanelli, and D. Gruber, "Toxicity of cobalt to freshwater indicator species as a function of water hardness," Aquatic Toxicology, vol. 22, no. 3, pp. 163–180, 1992.
22. J. M. Besser and K. J. Leib, "Modeling frequency of occurrence of toxic concentrations of zinc and copper in the upper Animas River," in Proceedings of the Technical Meeting, U.S. Geological Survey Toxic Substances Hydrology Program, D. W. Morganwalp and H. T. Buxton, Eds., vol. 1 of Contamination from Hardrock Mining: U.S. Geological Survey Water-Resources Investigations Report 99-4018A, pp. 75–81, Charleston, SC, USA, March 1999.
23. J. A. Erdman, L. A. Vradenburg, and S. C. Smith, "Extent of mine contamination plume on the Willow Creek floodplain, Creede, Colorado, as determined by willow leaf analysis," open-file report 2005-1267, U.S. Geological Survey, Reston, Va, USA, 2005, http://pubs.usgs.gov/of/2005/1267/pdf/OFR-1267.pdf.
24. M. Fleischer, A. F. Sarofim, and D. W. Fassett, "Environmental impact of cadmium: a review by the panel on hazardous trace substances," Environmental Health Perspectives, vol. 7, pp. 253–323, 1974.
25. U.S. Environmental Protection Agency (USEPA), "Quality criteria of water—1986," EPA 440/5-86-110, U.S. Environmental Protection Agency, Office of Water Regulations and Standards, Washington, DC, USA, 1986.
26. U.S. Environmental Protection Agency (USEPA), "Ambient water quality for aluminum," EPA 440/5-88-008, U.S. Environmental Protection Agency, Washington, DC, USA, 1988.

27. K. V. Beregeri and H. S. Patil, "The influence of pH on toxicity and accumulation of zinc in the fresh water fish Lepidocephalichtys guneta," Pollen Research, vol. 5, pp. 147–151, 1986.
28. A. M. Hilmy, N. A. El Domiaty, A. Y. Daabees, and H. A. Abdel Latife, "Some physiological and biochemical indices of zinc toxicity in two freshwater fishes, Clarias lazera and Tilapia zilli," Comparative Biochemistry and Physiology, vol. 87, no. 2, pp. 297–301, 1987.
29. F. Gostomski, "The toxicity of aluminum to aquatic species in the US," Environmental Geochemistry and Health, vol. 12, no. 1-2, pp. 51–54, 1990.
30. N. E. Kemble, W. G. Brumbaugh, E. L. Brunson et al., "Toxicity of metal-contaminated sediments from the upper Clark Fork River, Montana, to aquatic invertebrates and fish in laboratory exposures,"Environmental Toxicology and Chemistry, vol. 13, no. 12, pp. 1985–1997, 1994.
31. D. F. Woodward, W. G. Brumbaugh, A. J. DeLonay, E. E. Little, and C. E. Smith, "Effects on rainbow trout fry of a metals-contaminated diet of benthic invertebrates from the Clark Fork River, Montana,"Transactions of the American Fisheries Society, vol. 123, no. 1, pp. 51–62, 1994.
32. W. G. Wright, "Natural and mining-related sources of dissolved minerals during low flow in the upper Animas River basin, southwest Colorado," U.S. Geological Survey, Fact Sheet FS-148-97, 1997.
33. D. W. Nimmo, C. J. Castle, and J. M. Besser, "A toxicological reconnaissance of the upper Animas River watershed near Silverton, Colorado," in U.S. Geological Survey Open-File Report 98-297, D. S. Nimick and P. von Guerard, Eds., p. 19, 1998.
34. A. M. Farage, D. F. Woodward, D. Skaar, and W. G. Brumbaugh, "Characterizing the aquatic health in the Boulder River watershed, Montana," U.S. Geological Survey Water Resources Investigations Report 4018A, 1999.
35. J. M. Besser, W. G. Brumbaugh, T. W. May, S. E. Church, and B. A. Kimball, "Bioavailability of metals in stream food webs and hazards to brook trout (Salvelinus fontinalis) in the upper Animas River watershed, Colorado," Archives of Environmental Contamination and Toxicology, vol. 40, no. 1, pp. 48–59, 2001.

36. K. Walton-Day, J. L. Flynn, B. A. Kimball, and R. L. Runkel, "Mass loading of selected major and trace elements in Lake Fork Creek near Leadville, Colorado," U.S. Geological Survey, Scientific Investigations Report 2005-5151, 2001.
37. L. E. Schemel, B. A. Kimball, R. L. Runkel, and M. H. Cox, "Formation of mixed Al-Fe colloidal sorbent and dissolved-colloidal partitioning of Cu and Zn in the Cement Creek—Animas River Confluence, Silverton, Colorado," Applied Geochemistry, vol. 22, no. 7, pp. 1467–1484, 2007.
38. B. A. Kimball, R. L. Runkel, K. Walton-Day, and B. K. Stover, "Evaluation of metal loading to streams near Creede, Colorado," Scientific Investigations Report 2004-5143, U.S. Geological Survey, 2000.
39. Willow Creek Reclamation Committee (WRC), "Report on surface and mine water sampling and monitoring in Willow Creek Watershed, Mineral County, CO. (1999–2002)," 2004,http://www.willowcreede.org/waterquality/Final%20Surface%20Water%20Report.pdf.
40. Willow Creek Reclamation Committee (WRC), "Report on characterization of groundwater in the alluvial deposits beneath the floodplain of Willow Creek below Creede," 2004,http://www.willowcreede.org/waterquality/Complete%20Well%20Report.pdf
41. U.S. Environmental Protection Agency (USEPA), "Aquatic resources assessment of the Willow Creek Watershed," prepared by Ecosystems Protection Program U.S. Environmental Protection Agency, Region 8, Denver, Colorado for the Willow Creek Reclamation Committee Creede, Colorado, 2005,http://www.willowcreede.org/waterquality/WC_Assessment_110105.pdf.
42. R. L. McLaughlin, "Nelson Tunnel water management feasibility study for Willow Creek Reclamation Committee Creede, Colorado," McLaughlin Rincon Engineering Design Consulting, 2006,http://www.willowcreede.org/waterquality/WCRCNelsonTunnelTreatmentStudy.pdf.
43. S. Yochum, C. Villa, R. Brown, et al., "Willow Creek Stream restoration: planning report," USDA-NRCS, April 2007, http://www.sywater.org/files/WillowCreekStreamRestoration_PlanningReport.pdf.

44. T. A. Steven and J. C. Ratte, "Geology and structural control of ore deposition in the Creede District, San Juan Mountains, Colorado," USGS Professional Paper 487, 1965.
45. J. Crepin and R. L. Johnson, "Soil sampling for environmental assessment," in Soil Sampling and Methods of Analysis, M. R. Carter, Ed., pp. 5–24, Canadian Society of Soil Science, CRC Press, Boca Raton, Fla, USA, 1993.
46. Soil Survey Division Staff, "Soil survey manual," USDA-NRCS, U.S. Government Printing Office, 1993.
47. Soil Survey Staff and R. Burt, "Soil survey laboratory methods manual," version 4.0, USDA-NRCS, Soil Survey Investigations Report No. 42, U.S. Government Printing Office, Washington, DC, USA, 2004,http://soils.usda.gov/technical/lmm/.
48. J. K. Taylor, Quality Assurance of Chemical Measurements, Lewis Publishing, Chelsea, Mich, USA, 1988.
49. R. Burt, M. A. Wilson, M. D. Mays et al., "Trace and major elemental analysis applications in the U.S. cooperative soil survey program," Communications in Soil Science and Plant Analysis, vol. 31, pp. 1757–1771, 2000.
50. M. A. Wilson, R. Burt, and M. D. Mays, "Application of trace elements in the U.S. Cooperative Soil Survey Program," in Proceedings of the 6th International Conference of the Biogeochemistry ofTrace Elements (ICOBTE '01), p. 324, Ontario, Canada, 2001.
51. R. K. Shaw, M. A. Wilson, L. Reinhardt, and J. Isleib, "Geochemistry of artifactual coarse fragment types from selected New York City soils," in Proceedings of the 19th World Congress of Soil Science, Brisbane, Australia, August 2010.
52. F. D. Wilde, D. B. Radtke, J. Gibs, and R. T. Iwatsubo, Collection of water samples: U.S. Geological survey techniques of water resources investigations, book 9, chapter A4, 1999,http://pubs.water.usgs.gov/.
53. E. J. Velthorst, "Water analysis," in Manual for Soil and Water Analysis, P. Buurman, B. van Lagen, and E. J. Velthorst, Eds., pp. 121–242, Backhuys, Leiden, The Netherlands, 1996.
54. Soil Survey Staff, Keys to Soil Taxonomy, USD-NRCS, U.S. Government Printing Office, Washington, DC, USA, 11th edition, 2010.

55. P. J. Schoeneberger, D. A. Wysocki, E. C. Benham, and W. D. Broderson, Field Book for Describing and Sampling Soils, Version 2.0, USDA-NRCS, National Soil Survey Center, Lincoln, Neb, USA, 2002.
56. S. L. Lane, S. Flanagan, and F. D. Wilde, "Selection of equipment for water sampling," version 2.0, U.S. Geological Survey Techniques for Water-Resources Investigations, book 9, chapter A2, 2003,http://pubs.water.usgs.gov/twri9A2/.
57. F. D. Wilde, "Cleaning of equipment of water sampling," version 2.0, U.S. Geological Survey Techniques for Water-Resources Investigations, book 9, chapter A#, 2004, http://pubs.water.usgs.gov/twri9A3/.
58. U.S. Geological Survey, "Collection of water samples," version 2.0, U.S. Geological Survey Techniques for Water-Resources Investigations, book 9, chapter A4, 2006, http://pubs.water.usgs.gov/twri9A4/.
59. C. Kabala and B. R. Singh, "Fractionation and mobility of copper, lead, and zinc in soil profiles in the vicinity of a copper smelter," Journal of Environmental Quality, vol. 30, no. 2, pp. 485–492, 2001.
60. R. G. Wetzel, Limnology, WB Saunders, Philadelphia, Pa, USA, 1975.
61. U.S. Department of Agriculture, Natural Resources Conservation Service (USDA-NRCS), "Stream corridor restoration," U.S. Department of Commerce, National Technical Information Service, 2005.
62. E. R. Weiner, Applications of Environmental Aquatic Chemistry: A Practical Guide, CRC Press, New York, NY, USA, 2008.
63. C. Guéguen and J. Dominik, "Partitioning of trace metals between particulate, colloidal and truly dissolved fractions in a polluted river: the Upper Vistula River (Poland)," Applied Geochemistry, vol. 18, no. 3, pp. 457–470, 2003.
64. L. Sigg, H. Xue, D. Kistler, and R. Schönenberger, "Size fractionation (dissolved, colloidal and particulate) of trace metals in the Thur river, Switzerland," Aquatic Geochemistry, vol. 6, no. 4, pp. 413–434, 2000.

65. K. J. Leib, M. A. Mast, and W. G. Wright, "Characterization of selected water-quality constituents and seasonal loadings in the upper Animas River Basin, southwestern Colorado," U.S. Geological Survey, Water Resources Investigations Report, 2000.
66. B. A. Kimball, K. E. Bencala, and R. E. Broshears, "Geochemical processes in the context of hydrologic transport: reactions of metals in St. Kevin Gulch, Colorado," in Toxic Substances in the Hydrologic Sciences, A. Dutton, Ed., pp. 80–94, The American Institute of Hydrology, Minneapolis, Minn, USA, 1994.
67. B. A. Kimball, D. A. Nimick, L. J. Gerner, and R. L. Runkel, "Quantification of metal loading in Fisher Creek by tracer injection and synoptic sampling, Park County, Montana," U.S. Geological Survey Water-Resources Investigations Report 99-4119, 1999.
68. B. A. Kimball, R. L. Runkel, and L. J. Gerner, "Quantification of metal loading in French Gulch, Summit County, Colorado using a tracer-injection study," U.S. Geological Survey Water-Resources Investigations Report 98-4078, 1999.
69. B. A. Kimball, R. L. Runkel, and L. J. Gerner, "Quantification of mine-drainage inflows to Little Cottonwood Creek, Utah, using a tracer-injection and synoptic-sampling study," Environmental Geology, vol. 40, no. 11-12, pp. 1390–1404, 2001.
70. B. A. Kimball, R. L. Runkel, K. Walton-Day, and K. E. Bencala, "Assessment of metal loads in watersheds affected by acid mine drainage by using tracer injection and synoptic sampling: Cement Creek, Colorado, USA," Applied Geochemistry, vol. 17, no. 9, pp. 1183–1207, 2002.
71. T. E. Cleasby, D. A. Nimick, and B. A. Kimball, "Quantification of metal loads by tracer-injection and synoptic sampling methods in Cataract Creek, Jefferson County, Montana, August 1997," U.S. Geological Survey Water-Resources Investigations Report 00-4237, 2000.
72. A. Düker, A. Ledin, S. Karlsson, and B. Allard, "Adsorption of zinc on colloidal (hydr)oxides of Si, Al and Fe in the presence of a fulvic acid," Applied Geochemistry, vol. 10, no. 2, pp. 197–205, 1995.
73. J. G. Webster, P. J. Swedlund, and K. S. Webster, "Trace metal adsorption onto an acid mine drainage iron(III) oxy hydroxy

sulfate," Environmental Science and Technology, vol. 32, no. 10, pp. 1361–1368, 1998.

74. M. M. Benjamin and B. D. Honeyman, "Trace metals," in Global Biochemical Cycles, S. S. Butcher, R. J. Carlson, G. H. Orans, and G. V. Wolfe, Eds., pp. 317–352, Academic Press, New York, NY, USA, 1992.
75. R. P. Gambrell, J. B. Wiesepape, W. H. Patrick Jr., and M. C. Duff, "The effects of pH, redox, and salinity on metal release from a contaminated sediment," Water, Air, and Soil Pollution, vol. 57-58, pp. 359–367, 1991.
76. U.S. Department of Agriculture, Forest Service (USFS), "North Santiam River Turbidity Study," 1998.
77. S. Fendorf, G. Brown, and A. Spormann, "Researchers probe how microbes speed up acid production at mining sites," ScienceDaily, December 2003.
78. C. W. Childs, R. L. Parfitt, and R. Lee, "Movement of aluminium as an inorganic complex in some podzolised soils, New Zealand," Geoderma, vol. 29, no. 2, pp. 139–155, 1983.
79. U. Schwertmann, "The effect of pedogenic environments on iron oxide minerals," Advances in Soil Sciences, vol. 1, pp. 172–200, 1985.
80. R. L. Parfitt and C. W. Childs, "Estimation of forms of Fe and Al: a review, and analysis of contrasting soils by dissolution and Moessbauer methods," Australian Journal of Soil Research, vol. 26, no. 1, pp. 121–144, 1988.
81. P. W. Birkeland, R. M. Burke, and J. B. Benedict, "Pedogenic gradients for iron and aluminum accumulation and phosphorus depletion in arctic and alpine soils as a function of time and climate,"Quaternary Research, vol. 32, no. 2, pp. 193–204, 1989.
82. K. Wada, "Allophane and imogolite," in Minerals in Soil Environment, J. B. Dixon and S. B. Weed, Eds., Soil Science Society of America Book Series No. 1, pp. 1051–1087, 2nd edition, 1989.
83. J. B. Fellman, E. Hood, R. T. Edwards, and D. V. D'Amore, "Changes in the concentration, biodegradability, and fluorescent properties of dissolved organic matter during stormflows in coastal temperate watersheds," Journal of Geophysical Research G, vol. 114, no. 1, pp. 1–14, 2009.

84. D. Zabowski, C. L. Henry, Z. Zheng, and X. Zhang, "Mining impacts on trace metal content of water, soil, and stream sediments in the Hei River basin, China," Water, Air, and Soil Pollution, vol. 131, pp. 261–273, 2001.

85. A. Karczewska, "Metal species distribution in top- and sub-soil in an area affected by copper smelter emissions," Applied Geochemistry, vol. 11, no. 1-2, pp. 35–42, 1996.

86. Y. B. Mao and N. C. Uren, "Application of a new fractionation scheme for heavy metals in soils,"Communications in Soil Science and Plant Analysis, vol. 26, pp. 259–264, 1995.

87. W. Wilcke, S. Müller, N. Kanchanakool, and W. Zech, "Urban soil contamination in Bangkok: heavy metal and aluminium partitioning in topsoils," Geoderma, vol. 86, no. 3-4, pp. 211–228, 1998.

88. S. L. McGowen and N. T. Basta, "Heavy metal solubility and transport in soils contaminated by mining and smelting," in Heavy Metal Release in Soils, H. Magdi Selim and D. L. Sparks, Eds., chapter 4, pp. 89–107, CRC Press, New York, NY, USA, 2001.

89. I. Pais and J. B. Jones Jr., The Handbook of Trace Elements, St. Lucie Press, Boca Raton, Fla, USA, 1997.

90. F. Stevenson and M. S. Ardakani, "Organic matter reactions involving micronutrients in soils," inMicronutrients in Agriculture, pp. 36–58, Soil Science Society of America, Madison, Wis, USA, 1972.

91. A. Cottenie, Trace Elements in Agriculture and in the Environment, Laboratory of Analytical and Agrochemistry, State University Ghent, Ghent, Belgium, 1981.

92. U. Forstner, "Land contamination by metals: global scope and magnitude of problem," in Metal Speciation and Contamination of Soil, H. E. Allen, C. P. Huang, G. W. Bailey, and A. R. Bowers, Eds., pp. 1–33, CRC Press Lewis Publishing, 1995.

93. U.S. Forest Service (USFS), "Valles Caldera National Preserve, Soils, Existing Condition Report," Valles Caldera Trust, State of New Mexico, Sandoval and Rio Arriba Counties, USFS, 2007,http://www.fs.fed.us/r3/sfe/jemez_mtn_rest/reports/VCNP_ExistingConditions_Soils.pdf.

94. R. D. Knight and P. J. Henderson, "Smelter dust in humus around Rouyn-Noranda, Québec,"Geochemistry: Exploration, Environment, Analysis, vol. 6, no. 2-3, pp. 203–214, 2006.
95. F. N. Ponnamperuma, "The chemistry of submerged soils," Advances in Agronomy, vol. 24, pp. 29–96, 1972.
96. U. Förstner, Interaction at the Soil Colloid—Soil Solution Interface, NATO ASI Series, Kluwer Academic, Dodrecht, the Netherlands, 1991.
97. S. Satawathananont, W. H. Patrick Jr., and N. van Breemen, "Changes with time of redox potential, pH and chemical constituents of acid sulphate soils of Thailand," in Rice Production on Acid Soils of the Tropics, P. Deturck and F. N. Ponnamperuma, Eds., pp. 59–64, Institute of Fundamental Studies, Kandy, Sri Lanka, 1991.
98. W. Calmano, J. Hong, and U. Förstner, "Binding and mobilization of heavy metals in contaminated sediments affected by pH and redox potential," Water Science and Technology, vol. 28, no. 8-9, pp. 223–235, 1993.
99. R. Charlatchka and P. Cambier, "Influence of reducing conditions on solubility of trace metals in contaminated soils," Water, Air, and Soil Pollution, vol. 118, no. 1-2, pp. 143–167, 2000.
100. R. P. Narwal, B. R. Singh, and B. Salbu, "Association of cadmium, zinc, copper, and nickel with components in naturally heavy metal-rich soils studied by parallel and sequential extractions,"Communications in Soil Science and Plant Analysis, vol. 30, no. 7-8, pp. 1209–1230, 1999.
101. A. Vaněk, L. Borůvka, O. Drábek, M. Mihaljevič, and M. Komárek, "Mobility of lead, zinc and cadmium in alluvial soils heavily polluted by smelting industry," Plant, Soil and Environment, vol. 51, no. 7, pp. 316–321, 2005.
102. K. A. Yusuf, "Sequential extraction of lead, copper, cadmium and zinc in soils near Ojota waste site,"Journal of Agronomy, vol. 6, no. 2, pp. 331–337, 2007.
103. A. Kabata-Pendias and H. Pendias, Trace Elements in Soils and Plants, CRC Press, Boca Raton, Fla, USA, 2nd edition, 1992.
104. A. Chlopecka, "Forms of trace metals from inorganic sources in soils and amounts found in spring barley," Water, Air, and Soil Pollution, vol. 69, no. 1-2, pp. 127–134, 1993.

105. L. Ramos, L. M. Hernandez, and M. J. Gonzalez, "Sequential fractionation of copper, lead, cadmium and zinc in soils from or near Donana National Park," Journal of Environmental Quality, vol. 23, no. 1, pp. 50–57, 1994.
106. V. Matera, I. Le Hécho, A. Laboudigue, P. Thomas, S. Tellier, and M. Astruc, "A methodological approach for the identification of arsenic bearing phases in polluted soils," Environmental Pollution, vol. 126, no. 1, pp. 51–64, 2003.
107. M. Filippi, V. Golia, and Z. Pertold, "Arsenic in contaminated soils and anthropogenic deposits at the Mokrsko, Roundy, and Kasperske Hory gold deposits, Bohemiam Massif (CZ)," Environmental Geology, vol. 45, pp. 716–730, 2004.
108. A. K. Ghosh, D. Sarkar, D. Nayak, and P. Bhattacharyya, "Assessment of a sequential extraction procedure for fractionation of soil arsenic in contaminated soils," Archives of Agronomy and Soil Science, vol. 50, pp. 583–591, 2004.
109. L. Beesley, E. Moreno-Jiménez, R. Clemente, N. Lepp, and N. Dickinson, "Mobility of arsenic, cadmium and zinc in a multi-element contaminated soil profile assessed by in-situ soil pore water sampling, column leaching and sequential extraction," Environmental Pollution, vol. 158, no. 1, pp. 155–160, 2010.
110. R. Burt, J. V. Chiaretti, and D. J. Prevost, "Trace elements of selected soils in western Nevada and eastern California," Soil Survey Horizons, vol. 46, pp. 120–131, 2005.

Chapter 5

Multivariate Analysis of Metal Levels in Paddy Soil, Rice Plants, and Rice Grains: A Case Study from Shakargarh, Pakistan

Saadia R. Tariq and Nabila Rashid

Department of Chemistry, Lahore College for Women University, 45600 Lahore, Pakistan

ABSTRACT

The present study aims at determining the relationship between trace metal levels in paddy soils, rice plants, and rice grains obtained from these plants. The levels of selected metals (Fe, Co, Ni, Cd, Pb, and Cr) were determined by atomic absorption spectrophotometry in the soil, rice plants, and rice grain samples collected from paddy fields. All the metals were present at enhanced levels in paddy soil. Among the selected metals, Fe, Ni, Cd, and Cr were predominantly associated with oxidizable fraction. The metals such as Cr, Ni, Co, and Fe were significantly positively correlated in soil and plants, but no such correlation was observed in soil-grain matrix evidencing that these metals have a soil-based origin in the plants, but they were not translocated to grains. The Pb content of soil was strongly positively correlated with plants as well as the grains. The principal component analysis and cluster analysis were used to depict the origin of enhanced metal levels in rice plants. Under the given field conditions, different metals possess different translocation behaviours from soil to roots to shoots to grains. There is a dire need to implement the strategies for wise and optimum use of agrochemicals.

INTRODUCTION

The greater urgency to obtain more and more crop yields per capita has led to the excessive use of agrochemicals which not only supply the nutrients to the soils and adjust their pH but also protect the precious crops from various pests [1]. Despite all these benefits, the excessive and continuous use of agrochemicals may lead to a substantial increase in concentrations of major elements like K, Ca, and Mg and heavy metals such as Cd and Zn in soil solutions, thereby deteriorating the soil quality with a subsequent effect on plants that show stunted growth and thus lower crop yields [2].

In developing countries, the heavy-metal-contaminated sewage sludges are also used as fertilizers, which on continuous application may lead to the accumulation of toxic metals into the soil. The subsequent uptake of metals in the edible parts of vegetative tissues may result in a direct pathway into the human food chain [3]. Moreover, these metals may leach from contaminated soils into the water bodies thereby deteriorating their quality as well [4]. Therefore, attention has been diverted towards the determination of metal levels in various agricultural soils and their extent of bioavailability to plants due to their detrimental effects on soil flora and fauna and their potential human health risks [5–10].

Rice is a favorite food for more than three billion people around the world [11]. In Pakistan, rice is grown under diverse climatic and edaphic conditions. Paddy soils are specifically of interest because for the cultivation of rice; these soils are kept submerged for longer periods of time, which may cause reduction and subsequent mobilization of various metals such as Fe and Mn [12]. Therefore, paddy soils have been of interest to the scientists for the determination of the levels and sources of heavy metals [1]. Qishlaqi and Moore reported fungicides as major cause of Cu contamination of agricultural soils [7]. In rice-consuming countries like Iran, rice was suggested as one of the major sources of Cd intake due to high potential of these plants to accumulate Cd from contaminated soils into their grains [13].

In view of the hazards caused by the accumulation of heavy metals in the soil due to excessive agrochemical usage, the present study was formulated to determine the metal contents of soil samples collected from paddy fields of Tehsil Shakargarh, Pakistan, to determine the fractionation of metals present in paddy soils to study their bioavailability and mobility, and then to study the uptake of metals by rice plants and their translocation to edible parts. The study will not only highlight the metal contamination of the paddy fields arising due to excessive and unwise use of agrochemicals but also help to identify the potential risk of elevated levels of these metals to the consumers.

EXPERIMENTAL METHODOLOGY

Description of Sampling Site

The samples were collected from paddy fields of tehsil Shakargarh of Punjab province of Pakistan. The city is located at 32°16′0N 75°10′0E on the west bank of the river Ravi, encompassing an area of about 1272 km^2. Its rich and fertile lands are well known for wheat and top-quality rice cultivation. The main sources of irrigation in the sampling area are tube wells and electric pumps (water table 40–50 ft). The soil texture here is loamy, which is considered to be best for agricultural purposes because this can be ploughed easily and has tendency to retain good moisture content. Flour mills and rice processing units are the only industries present here.

Sample Collection and Pretreatment

A total of 35 samples each of top soil (1–5 cm), rice plants, and rice grains were collected in triplicate in 500 g capacity from different rice fields located at a distance of 1 km apart. The samples were properly labeled and immediately transferred to laboratory for analysis.

The collected soil samples were cleaned manually for twigs, roots, stones, or any other foreign matter, homogenised, and dried in an electric oven set at 80°C for 24 h. The rice plant samples were thoroughly washed with deionised water, dried in shade, chopped, homogenized, and stored in zip mouthed bags. Similarly, the rice grains after cleaning, washing, and grinding were stored in polythene bottles.

Analytical grade reagents with a certified purity of 99.9%, procured from E. Merck (Germany), were used throughout the work. The stock solutions of metals (1000 ppm) contained in strong opaque polyethylene bottles were also purchased from E-Merck Germany and stored in a cool and dry place.

Determination of Soil Physicochemical Parameters

The physicochemical parameters of the soil samples were determined by using the water extracts of these samples, prepared by shaking soil sample with deionized, distilled water in 1 : 1 ratio for 5 minutes. Subsequently, the contents were equilibrated for thirty minutes and filtered. The filtrate was used for the determination of physicochemical parameters in accordance with standard methods. The moisture content of the soil samples was determined gravimetrically [14].

Determination of Metal Contents of Soil Samples

In order to determine the metal contents of soil samples, air-dried and sieved soil samples, (2 g) were digested by using HNO_3 and H_2O_2 [14]. The digested samples were then diluted with distilled water and analyzed for the selected metals by using Hitachi Flame Atomic Absorption Spectrophotometer (FAAS) model Z-5000.

The speciation studies of selected metals in the soil matrix were carried out by adopting the sequential extraction procedure as recommended by Community Bureau of References [15]. The procedure comprised of extracting exchangeable, reducible, oxidizable, and residual fractions by using glacial acetic acid, hydroxyl ammonium chloride, H_2O_2-ammonium acetate, and nitric acid, respectively. After each extraction, the contents were centrifuged and subsequently filtered. All the four fractions obtained were analyzed for the estimation of selected metals, that is, Co, Fe, Cd, Ni, Pb, and Cr by FAAS.

For the determination of trace metals in the rice grains and plant samples, wet acid digestion method was used whereby one gram portion of sample was digested with $HClO_3$ and conc. HNO_3 in the presence of traces of H_2SO_4. The contents were cooled to room temperature and diluted with doubly distilled water. For Fe and Mn determination, the contents were boiled for few minutes before filtering [14]. All the analyses were performed in triplicate and the corresponding result matched within 0.05%.

Statistical Analysis of Data

In order to develop the better understanding of the obtained data set on various physicochemical parameters of soil and various metals present in the soil, plants, and grains samples, the univariant and multivariant statistical analyses were performed. They involved the extraction of basic statistical parameters like mean, standard deviation, skewness, and kurtosis essential to determine the spread and distribution of measured data. Correlation coefficient analysis was carried out to establish the correlation pattern of various metal pairs in the same media and also in different media like soil, plant, and grain samples. The multivariate statistical analysis in terms of principal component analysis and cluster analysis were performed by using statistica software [16] to get a better understanding of sources of various metals in the soil, plant, and grain samples [17].

The mobility index values of the soil samples were determined by using the formula provided in literature [18]. A high mobility index of a metal was indicative of its bioavailable and mobile nature [19].

RESULTS AND DISCUSSION

Physicochemical Parameters

The physicochemical parameters of thirty-five soil samples collected from different paddy fields of Shakargarh, Pakistan are provided in Table 1. The pH of collected samples was found to be alkaline with small variations. This alkaline pH may decrease the metal mobility by the formation of precipitates, by increasing the number of available adsorption sites and decreasing the competition of H^+ for adsorption, and also by increasing the metal stability with humic substances [20]. The high moisture contents were noted for these soil samples that are characteristic of loamy textured soil.

Table 1: Descriptive statistics for physico-chemical parameters of soil samples (n = 3 5)

	Moisture content (%)	pH	Conductivity (µS/cm)	Cl– (mg/kg)	NO3- (mg/kg)	SO42- (mg/kg)
Minimum	8.000	8.1	778.0	2120	1834	8200
Maximum	17.00	8.4	1090	6.24 × 104	1.440 × 105	2.30 × 104
Mean	11.79	8.2	975.4	9828	2.43 × 104	1.401 × 104
Median	12.00	8.2	1006	3160	1.485 × 104	1.38 × 104
S D	2.467	0.081	90.23	1.77 × 104	3.389 × 104	4819
S.E	0.504	0.017	18.41	3629	6917. 961	983.7
Kurtosis	−0.623	0.070	−0.022	4.597	9.297	−1.121
Skewness	0.546	0.975	−1.121	2.465	3.191	0.374
Variance	6.085	0.007	8141	3.16 × 108	1.14 × 109	2.32 × 107

The mean conductivity of the samples was found to be 975.4 μScm^{-1}. The chloride, nitrate, and sulphate levels exhibited the order NO_3 >

$SO_4^{2-} > Cl^-$ with their mean concentrations standing at 24.30, 14.01, and 9.828 g/kg, respectively. Sources of these enhanced levels were traced in agronomic activities such as use of natural and synthetic fertilizers, herbicides, and pesticides used to improve the crop yields. Parental soil quality is another factor that may contribute to the enhanced levels of these anions.

Metal Levels in the Soil, Plants, and Grain Samples

The total metal contents of soil, plant, and grain samples are presented in Table 2. In the soil samples, Fe was found to be present at the highest mean concentration of 4074 mg/kg ranging from 1823 to 7498 mg/kg. It was followed by Pb standing at 1065 mg/kg with an observed standard deviation of 1092. The order for mean levels of the rest of the metals was Co > Ni > Cr > Cd with their respective levels being 747.2 > 85.84 > 37.80 > 35.02 mg/kg. The large differences observed between mean and median values evidenced nonnormal distribution of the data. Moreover, all the metals exhibited the levels far more exceeding the National Environmental Quality Standards as depicted in Figure 1 [21]. The data set obtained for metals presented an order quite similar to those observed in previous studies [2, 22]. Although Fe concentration levels are high in the studied soil samples as compared to other metals, these levels are lower than the maximum values established by the European and Spanish legislation for rice-forming soils [23].

Table 2: Descriptive statistics for metal levels (mg/kg) in soil, plants, and grains (n = 3 5)

	Fe	Co	Ni	Cd	Pb	Cr
Soil						
Min.	1823	13.76	60.02	1.025	113.8	13.76
Max.	7498	2898	99.67	78.90	2967	89.76
Mean	4074	747.2	85.84	35.02	1065	37.80
Median	3455	90.22	90.68	30.70	725.5	25.51
S D	2249	1180	13.61	33.95	1092	27.73
S.E	459.1	240.9	2.779	6.93	222.9	5.662

Kurtosis	–1.263	–0.497	–0.674	–2.107	–1.115	–0.600
Skew ness	0.591	1.240	–0.936	0.054	0.656	1.115
Plant						
Min.	639.3	61.97	49.11	0.040	31.34	4.280
Max.	1624	80.19	64.06	2.060	67.53	15.54
Mean	1256	71.58	54.50	0.849	48.81	9.173
Median	1382	72.08	52.41	0.648	48.19	8.435
S D	427.1	9.975	7.094	0.990	20.20	5.776
S.E	213.6	4.988	3.547	0.495	10.10	2.888
Kurtosis	2.922	–5.800	–0.284	–2.967	–5.925	–4.738
Skew ness	–1.557	–0.035	1.050	0.540	0.013	0.221
Grain						
Min.	243.9	40.49	38.59	0.220	38.63	4.653
Max.	1019	62.03	52.59	1.642	55.88	12.21
Mean	536.1	49.89	46.34	0.859	45.75	6.988
Median	440.8	48.51	47.09	0.788	44.25	5.544
S D	335.2	8.959	6.420	0.741	8.22	3.507
S.E	167.6	4.479	3.210	0.371	4.112	1.754
Kurtosis	2.845	1.789	–2.728	–5.202	–2.762	3.736
Skew ness	1.522	0.882	–0.424	0.139	0.561	1.911

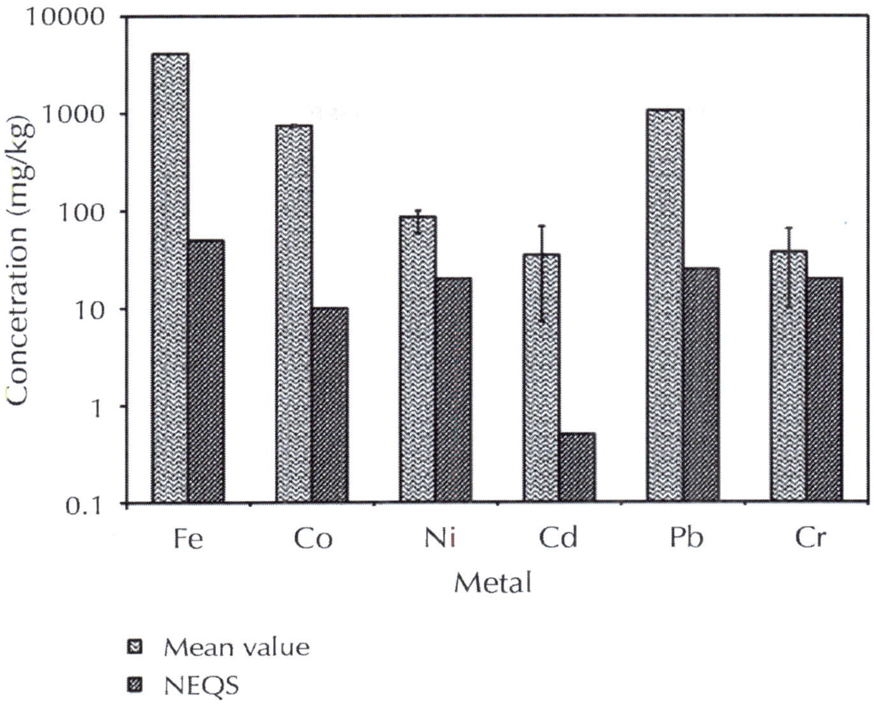

Figure 1: Comparison of mean metal levels in soil (±S.D.) with NEQS.

The addition of fertilizers/micronutrients to the agricultural soils may result in enhanced Fe and Mn levels [24]. These Fe rich soils, under aerobic conditions may act as As accumulators [25, 26], which is found in the form of arsenate bound to clay particles. A good correlation, therefore, has been reported between extractable Fe, Mn, and As [27, 28], which may lead to high plant exposure to these metals and thus risk of plant uptake with consequent contamination of whole food chain [29]. Rice has been found to be most efficient at accumulating As into grains than other plants but if rice is grown under aerobic conditions, the extent of bioaccumulation of As and other metals may be reduced [30]. Further, reducing conditions and presence of organic matter may promote as mobilization to ground waters through ion exchange mechanisms with phosphorous derived from fertilizers [31–33] thereby making it unfit for any viable human use.

In order to study the influence of enhanced metal levels in the soil on the plants and human populations, the metal content of rice plants and grains raised in these soils was also determined and incorporated in Table 2 in the form of descriptive statistics. The data evidenced Fe to be present at the highest mean level of 1256 mg/kg in rice plants ranging from 639.3 to 1624 mg/kg. It was followed by Co being present at mean level of 71.58 mg/kg. Ni and Pb were found to be present at mean concentrations of 54.50 and 48.81 mg/kg, respectively. Cr was present at quite smaller levels than Co and Ni but still at appreciable levels of 9.173 mg/kg. Cd was the metal that was present at the least mean concentration of 0.849 mg/kg. Thus, the overall order of mean concentrations of metals in rice plants was found to be Fe > Co > Ni > Pb > Cr > Cd.

The corresponding data set for rice grains obtained from these rice plants also presented Fe to be present at the highest mean concentration of 536.1 mg/kg. Co was the metal that secured the second highest concentration of 62.03 mg/kg against the mean concentration of 49.89 mg/kg. Pb was present at a mean concentration of 45.75 mg/kg, far exceeding the permissible limits in rice grains in China and Australia, that is, 0.3 to 5.2 mg/kg [34–36]. Cd was found at sub-ppm levels and was well within the safety limits that were set to range from 0.02 to 1 mg/kg in China, Australia, and Japan for rice grains [26]. The order of mean metal levels for the present study was found to be Fe > Co > Ni > Pb > Cr > Cd. Thus, the studied rice grains exhibited mean Pb levels far exceeding the safe limits. The data set obtained for grains also presented mean metal order quite comparable to that found in plants.

The orders of various metals observed in three media were quite comparable with few exceptions, that is, Pb secured the second highest position in soil while in the plant and grain samples it was present at the fourth position. The highest levels of Fe in the three media depicted their similar origin. The different orders of metals in these media may be explained on the basis of differences in nature of these metals and their bioavailability in the soil.

Ni, Co, and Cr concentrations were also comparable in both soil and plant samples with small variations as the bioavailability of metals depends upon the nature of the metal and the properties of soil. It is observed that metals from anthropogenic sources tend to be more mobile than pedogenic or lithogenic ones.

High concentration of metals in the soil does not necessarily imply their availability to plants. In the solid phase, the metals are distributed among the various soil components thereby producing various physicochemical forms that determine metal mobility [19]. Thus, in order to better assess the bioavailability of metals and their chemical association with the soil components metal fractionation behavior was studied by using the fractionation scheme proposed by community bureau of references.

The results of the fractionation analysis of the metals in the soil are presented in Table 3, where the mean values for each fraction are reported. The distribution of the metals in various operationally defined fractions evidenced that Fe was mostly found to be associated with oxidizable fraction (1743 mg/kg). The reducible fraction of this metal was also present at an appreciable concentration of 1590 mg/kg. The least amount was found to be associated with acid soluble fraction, that is, 80.77 mg/kg. For Co most of the metal was found to be associated with oxidizable fraction (372.5 mg/kg). The overall order of mean levels of all the fractions was oxidizable > residual > reducible > acid soluble. Ni being present at a total concentration of 85.84 mg/kg was mostly associated with oxidizable and residual fractions with mean concentration of 24.10 and 19.61 mg/kg, respectively. Reducible fraction of this metal was present at mean level of 16.67 mg/kg, while the acid soluble fraction was found to be present at mean levels of 14.86 mg/kg. The mobility index of this metal was found to be 0.183.

Table 3: Fractionation studies of metals in soil of rice fields, *n* = 3 5 (mean ± SD)

	aF1	bF2	cF3	dF4	Mobility index	Total metal
Fe	80.77±59.72	1590 ± 376.9	1743.2 ± 2272	198.40 ± 155.2	0.037	4073.5±2249
Co	53.10±60.60	70.87 ± 90.65	372.5 ± 618.4	117.50 ± 171.8	0.185	747.2±1179
Cr	5.552±2.174	5.215 ± 1.759	16.06 ± 19.51	6.885 ± 5.593	0.183	37.8±27.57

Ni	14.86±9.416	16.67 ± 9.173	24.10 ± 11.16	19.61 ± 14.06	0.272	85.84±13.28
Pb	391.2±477.3	216.1 ± 345.6	377.4 ± 381.8	18.26 ± 9.693	0.330	1065±1081
Cd	9.012±8.712	9.011 ± 8.794	10.80 ± 10.44	2.433 ± 2.046	0.219	35.02±33.78

[a]Acid soluble. [b]Reducible. [c]Oxidizable. [d]Residual.

Cd, another heavy metal, was present at a total mean concentration of 35.02 mg/kg. The highest concentration of this metal was found to be associated with oxidizable fraction, that is, 10.80 mg/kg. The acid soluble and reducible fractions were present at comparable levels of 9.012 and 9.011 mg/kg, respectively. Least concentration of the metal was found to be associated with residual fraction. The percent recovery of the metal was found to be 85.40 with a mobility index of 0.219.

Pb being present at a total concentration of 1065 mg/kg exhibited a total recovery of 99.60% during distribution among the four fractions. Its highest concentration was present in acid soluble fraction. Oxidizable fraction exhibited the second highest concentration of 377.4 mg/kg, while reducible fraction was present at about 216 mg/kg concentrations. The least concentration of this metal was found to be associated with residual fraction. The mobility index of the metal was found to be 0.330. Cr, one of the most hazardous metals, was found to be associated mostly with oxidizable fraction (16.06 mg/kg). The order of the rest of the fractions was found to be residual > acid soluble > reducible. The percentage recovery of this metal was found to be 96.52% with a mobility index of 0.183.

The mobility of an element is its capacity to pass into soil compartments where it is less energetically retained [37]. The extractable content of a heavy metal is an indicator of the quantity of metals available to plants. The proportion of this fraction relative to the total amount of the metal is also an indicator of its comparative mobility [4]. The comparison of mobility index values evidenced Pb to be highly mobile and thus bioavailable. Thus, its higher concentration may leach to water and enter the plant bodies. This fact justified the higher concentration of Pb in plants as well as grain samples. Next higher mobility index value was observed for Ni, which was also concentrated into the plant bodies and grain samples. The rest of the metals followed the order of mobility indices as follows: Cd > Co > Cr > Fe.

Correlation Coefficient Matrix

The deficiency or excess of micronutrients in soil is directly associated with the plant uptake, which is indirectly associated with soil physicochemical characteristics and plant species. The physicochemical characteristics of soil are known to control the fate of the metals and largely influence the plant-soil interaction through rhizospheric processes.

The correlation coefficient matrix for various metals and physicochemical parameters of soil is presented in Table 4. The pH of these soil samples was found to be significantly negatively correlated with Co, Pb, and Cr with r-values of 0.495, 0.410, and 0.483, respectively. Chloride was also found to be significantly negatively correlated with Fe and Ni. Sulphate on the other hand was found to be significantly positively correlated with Pb, Co, and Cr with r-values of 0.511, 0.696, and 0.617, respectively.

Table 4: Correlation coefficient matrix for various parameters in respective soil, plants, and grain samples (n = 3 5)

	Fe	Co	Ni	Cd	Pb	Cr
Moist. Content (%)	0.374	−0.392	−0.088	−0.159	−0.296	−0.325
pH	0.018	−0.495	−0.097	−0.259	−0.41	−0.483
EC (µS/cm)	0.008	0.366	−0.565	0.693	0.604	0.464
Cl− (mg/kg)	−0.39	−0.221	−0.713	0.318	0.106	−0.112
NO-3 (mg/kg)	−0.071	0.217	−0.163	0.332	0.35	0.246
SO42- (mg/kg)	−0.187	0.696	0.336	0.235	0.511	0.617
Soil						
Fe	1.000					
Co	0.127	1.000				
Ni	0.554	0.085	1.000			
Cd	−0.352	0.634	−0.686	1.000		

Pb	−0.155	0.869	−0.372	0.912	1.000	
Cr	0.050	0.978	−0.080	0.758	0.930	1.000
Plant						
Fe	1.000					
Co	0.260	1.000				
Ni	0.155	−0.914	1.000			
Cd	−0.632	−0.913	0.668	1.000		
Pb	−0.385	−0.991	0.852	0.959	1.000	
Cr	−0.135	−0.992	0.958	0.853	0.966	1.000
Grains						
Fe	1.000					
Co	0.183	1.000				
Ni	−0.445	−0.930	1.000			
Cd	−0.314	0.743	−0.446	1.000		
Pb	−0.593	−0.563	0.820	0.132	1.000	
Cr	−0.322	−0.164	0.459	0.474	0.860	1.000

r- Values are significant at ±0.282 at $P = 0.05$

The data presented in Table 4 evidenced a number of strong correlations among various metal pairs in different media. In case of soil, a strong positive correlation was observed between Cd and Pb with a r-value of 0.912, and Pb and Cr with r-value of 0.930. Cr was also strongly positively correlated with Co (r-value 0.978). Similarly, Pb was strongly positively correlated with Co (0.869). The strong positive correlations evidenced the similar origin of these metal pairs. Three of the studied metal pairs in the soil were found to be significantly negatively correlated, that is, Cd–Fe, Ni–Cd, and Pb–Ni evidencing the increase in the concentration of one metal with the decrease in the concentration of another.

In plant samples, Co was found to be significantly negatively correlated with almost all the metals with the exception of Fe. Cd–Fe pair in plant samples was found to be negatively correlated, and Cd was positively correlated with Pb and Cr similar to soil. In case of grain samples, the strongest negative correlation was observed between Co–Ni pair with an r-value of 0.930. Fe was also found to be negatively correlated with Ni and Pb. Similarly, negative correlation was observed between Co–Pb pair. The strongest positive correlation was found in Pb–Cr pair, a situation quite similar to one encountered in plant and soil samples, evidencing their common origin in the three media. Ni–Cd pair in rice grains was significantly negatively correlated similar to soil. Cu–Ni and Co–Pb pairs were found to be significantly negatively correlated in plant and rice grain samples while they exhibited a significantly positive correlation in plant matrix, the reason may be traced in the phytoavailability of metals.

The correlation coefficient matrix for the metals in different media, that is, soil, plant, and grains is presented in Table 5. In case of soil versus plant metal correlation, the strongest positive correlation was observed between Cr contents in the two media, evidencing that the source of enhanced Cr levels in the plants originates from their enhanced levels in the soil. It was also supported by the high percentage of oxidizable fraction of Cr present in the soil. No such correlation was observed between Cr of soil and Cr of grains. It was indicative of the fact that although soil enhanced levels lead to absorption of Cr from soil, this excessive Cr was not translocated to the grains. It was also evidenced by a negative correlation between Cr (plants) and Cr (grains).

Table 5: Correlation coefficient matrix for metal levels in different media (n = 3 5)

	Fe(s)	Co(s)	Ni(s)	Cd(s)	Pb(s)	Cr(s)
Fe(p)	0.406	−0.240	0.342	−0.177	0.358	0.301
Co(p)	−0.182	0.488	−0.597	−0.498	−0.602	−0.437
Ni(p)	0.535	−0.507	0.557	0.412	0.487	0.604
Cd(p)	−0.031	−0.128	0.063	0.180	0.080	−0.149
Pb(p)	0.672	−0.799	0.815	0.696	0.604	0.889
Cr(p)	0.783	−0.809	0.792	0.653	0.646	0.904
	Fe(s)	Co(s)	Ni(s)	Cd(s)	Pb(s)	Cr(s)
Fe(g)	0.400	−0.209	0.048	0.565	−0.332	0.165
Co(g)	−0.535	0.566	−0.599	−0.195	−0.604	−0.378
Ni(g)	0.270	−0.222	0.348	0.256	0.596	0.444
Cd(g)	−0.457	0.481	−0.338	−0.429	−0.427	−0.323
Pb(g)	0.312	−0.407	0.631	0.059	0.672	0.463
Cr(g)	−0.259	0.082	0.166	0.078	0.130	0.129
	Fe(g)	Co(g)	Ni(g)	Cd(g)	Pb(g)	Cr(g)
Fe(p)	−0.066	0.076	0.237	0.221	0.161	0.013

Co(p)	−0.290	0.072	−0.353	0.114	−0.318	−0.500
Ni(p)	0.033	−0.030	0.415	0.170	0.250	0.242
Cd(p)	−0.211	0.399	−0.517	−0.146	−0.346	−0.398
Pb(p)	0.294	0.016	0.212	0.104	0.325	0.409
Cr(p)	−0.252	0.109	−0.428	0.094	0.049	−0.303

r- Values are significant at ±0.282 at $P = 0.05$ levels; s: soil, p: rice plant, and g: rice grain.

Pb of the soil was also significantly positively correlated with Pb of plants. Similarly Pb of soil was also significantly positively correlated with Pb contents of rice grains. As the grains constitute a direct edible part of plants, their accumulation of Pb is quite hazardous and may lead to various health hazards among the people consuming it. Similarly, Ni, Co, and Fe of soil were also found to be strongly positively associated with these metals in plant samples, but in case of rice grains only Ni was found to be significantly positively correlated. This matrix evidenced that enhanced levels of almost all the studied metals in the soil lead to an enhanced uptake by the arable parts of rice plants but these metals with the exception of Pb were not translocated to the grains, thus not posing any threat to human beings directly but indirectly they may constitute hazards because these arable parts may be used as food by the animals that constitute food for humans.

Source Apportionment Studies

Principle Component Analysis (PCA)

The sources of various metals in soil, plants, and grain samples were determined in terms of varimax normalized factor loading (Table 6). Only two rotated factors were obtained for the total metal contents of various metals present in thirty-five soil samples collected from different rice fields with a cumulative eigenvalue of 88.53%. The first factor received major contribution from Co, Cd, Pb, and Cr, while the second factor being associated with an eigen value of 1.746 received major loadings from Fe and Ni. It is noteworthy that the metals arising from nitrate fertilizers, that is, Cd and Co, are grouped together in a single factor depicting their similar source.

Table 6: Varimax normalized factor loadings for metals in soil, plant, and grain samples

Factor	Soil		Plant		Grain	
	1	2	1	2	1	2
Fe	0.069	0.834	0.413	0.808	– 0.7	−0.070
Co	0.912	0.18	−0.555	0.322	−0.086	– 0.809
Ni	−0.136	0.912	0.813	−0.042	0.266	0.809
Cd	0.787	−0.587	0.256	– 0.86	0.628	−0.657
Pb	0.949	−0.275	0.895	0.131	0.701	0.528
Cr	0.977	0.016	0.927	0.099	0.807	0.110
Eigenvalue	3.566	1.746	2.875	1.516	2.453	1.689
% Total variance	59.43	29.11	47.92	25.27	40.89	28.15
Cumulative eigenvalue	3.566	5.312	2.875	4.391	2.453	4.142
Cumulative %	59.43	88.53	47.92	73.19	40.89	69.04

The most significant sources of Fe in the soil are various fertilizers that are added to the soil to replenish various deficiency symptoms. It may also have geologic origins. It has been found that soils rich in Fe and Mn are also having enhanced As levels [38]. The sources of Pb in soil are traced in various fungicide and pesticide sprays that are used at various stages of crop production, containing Pb as an impurity [39]. It is known that phosphatic fertilizers are manufactured by using the phosphate ore, which is contaminated with different metals such as Pd, Cr, and Fe. These metal contaminants also become a part of fertilizer and on application to soil lead to hazardous effects [40]. This Pb contamination of paddy soils may change the soil microorganisms and soil enzymatic activities and deteriorate the soil fertility, which may directly affect the rice physiological indices and results in rice yield decline and quality deterioration. Ultimately, Pb accumulates in human body through food chain and thus endangers human health [34].

These fertilizers contain not only higher amounts of Fe, Co, and Pb but also traces of Cd and Ni as impurities that constitute the sources of their enhanced levels in the soil. The levels of heavy metals determined by several authors in phosphate fertilizers were 0.1–170 mg kg^{-1} Cd, 1–12 mg kg^{-1} Co, 7–38 mg kg^{-1} Ni, 7–225 mg kg^{-1} Pb while in the nitrate fertilizers they were 0.05–8.5 mg kg^{-1} Cd and 5.4–12 mg kg^{-1} Co. Moreover, the heavy metals also form a part of the active compounds of pesticides. Thus repeated and excessive fertilizer and pesticide applications may lead to much enhanced concentrations of heavy metals in the soil [22].

Sewage sludge is commonly used in agriculture as soil conditioner/fertilizer due to easy availability and low costs [41]. But it may also contain substantial amounts of toxic metals such as Zn, Cu, Cr, Ni, Cd, and Pb [42–44], which are quite mobile due to the presence of organic matter in sewage sludge [45]. In addition to direct application, municipal sewage sludge disposed of through land filling is also a potential source of spreading toxic metals to nearby lands [46]. Thus, sewage sludge is also a major contributor of metals in agricultural lands and poses serious threats to human health and environment [47, 48].

The varimax normalized factor loadings were also recorded for the metal data of rice plant samples to trace the sources of various metals. Similar to the soil samples again, two rotated factors were obtained. The first factor being associated with a cumulative variance of 47.92% received major contribution from Ni, Pb, and Cr, thereby depicting similar origin of these metals in the rice plants from agrochemicals that are used simultaneously and contain these metals. The second factor received major contribution from two metals only, that is, Fe and Cd.

The rice grains are the edible part of the plant used throughout the world, so it is quite imperative to evaluate their quality in terms of total metal content. Thus varimax normalized rotation of metal data of rice grains yielded two rotated factors with cumulative variance of 69.04%. First factor received major loading for Pb and Cr similar to plant samples. Fe was another major contributor of this factor. Factor 2 was mainly associated with Co and Ni with an eigenvalue of 1.698.

The food chain (soil-plant-human) pathway is recognized as one of the major pathways for human exposure to soil contamination. However, the ingested dose of heavy metals is not equal to the absorbed pollutant dose in reality, as a fraction of the ingested heavy metals may be excreted, with the remainder accumulated in body tissues where they affect human health [19]. Similarly, only limited quantities of metals are absorbed into the upper parts of the plants, probably because the roots act as a barrier to the translocation of metals within the plant [49]. The phenomenon is thus responsible for different metal levels in different parts of plants.

Cluster Analysis

The clustering behavior of the selected metals in soil, plant, and grain samples was studied by cluster analysis and is depicted in Figure 2. In case of soil samples, the selected metals formed two primary clusters within linkage distance of 0.5. One of the primary clusters comprised of Fe and Ni depicting their similar origin. The second cluster comprised of Cr, Pb, Cd, and Co, all linked within linkage distance of 0.2, evidencing their close association.

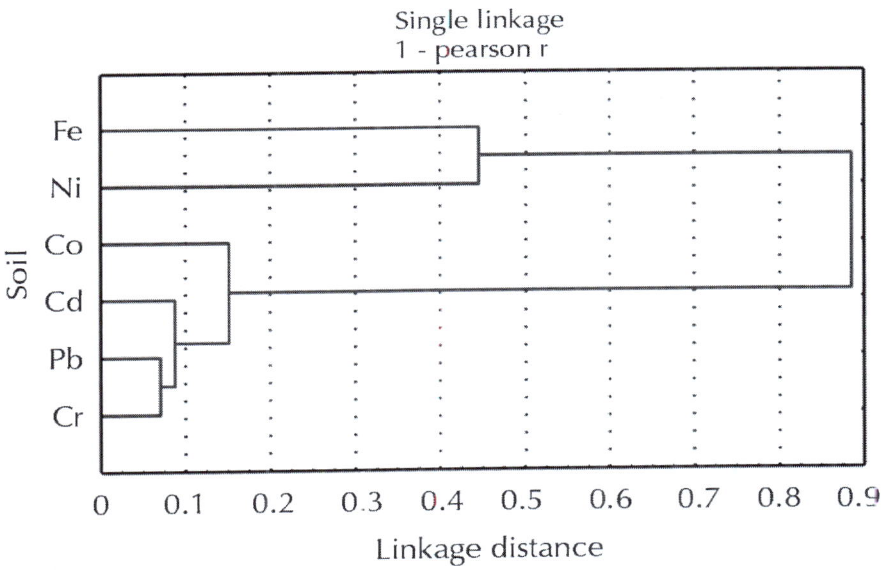

(a)

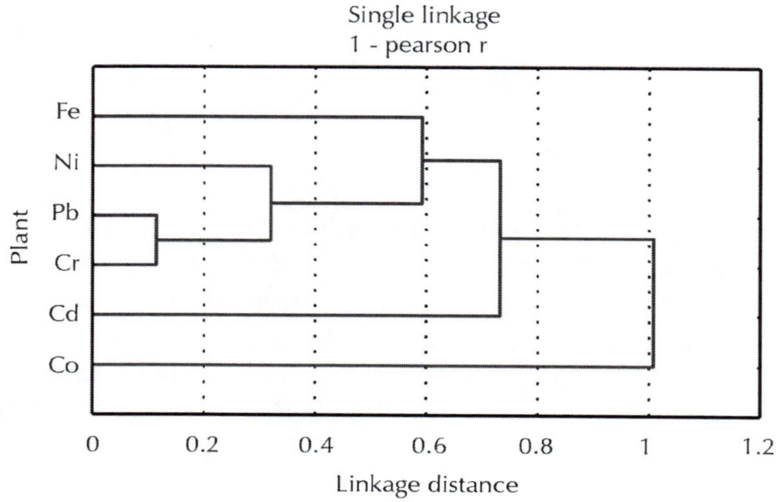

(b)

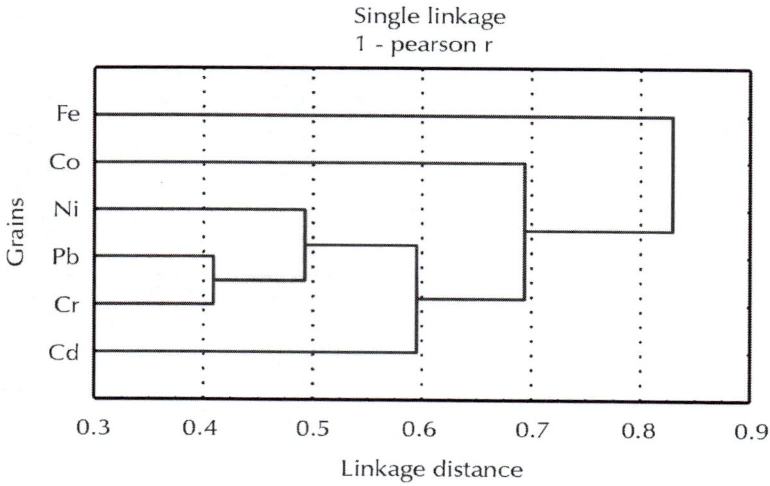

(c)

Figure 2: Cluster analysis of metals in soil, plants, and grains.

In case of plant samples, Fe stood alone depicting its different source. Similar to plant samples Pb, Cr, and Ni were found to be associated closely within linkage distance of 0.5, depicting their common origin in plants and grains. On overall basis, the cluster analysis confirmed the findings of the principle component analysis.

CONCLUSIONS

The present study evidenced enhanced levels of all the studied metals in the paddy soil samples due to abundant and unmanaged use of agrochemicals. The sequential extraction procedure revealed that the oxidizable and reducible phases of metals like Fe and Ni were predominant and they exhibited reduced mobility index values in the soil samples. Pb and Cd possessed high mobility due to their very high extractability in acid soluble fraction. The correlation coefficient matrix evidenced that enhanced metal levels in soil may lead to their accumulation in aerial parts of plants but not in edible parts of plants, that is, grains. A quite hazardous situation was observed where the rice grains in addition to aerial parts of plants also accumulated Pb, which may lead to various health hazards. Despite these hazards, rice plants may act hyper-accumulators of Cr, Ni, Co, and Fe by accumulating these metals in their aerial parts but not in edible parts. The study also evidenced that there is a dire need to address the problem of agricultural soil pollution and to educate the farmers about the optimum and wise use of agrochemicals.

REFERENCES

1. S. C. Wong, X. D. Li, G. Zhang, S. H. Qi, and Y. S. Min, "Heavy metals in agricultural soils of the Pearl River Delta, South China," Environmental Pollution, vol. 119, no. 1, pp. 33–44, 2002.
2. S. E. Lorenz, R. E. Hamon, S. P. McGrath, P. E. Holm, and T. H. Christensen, "Applications of fertilizer cations affect cadmium and zinc concentrations in soil solutions and uptake by plants," European Journal of Soil Science, vol. 45, no. 2, pp. 159–165, 1994.

3. K. C. Sekhar, K. R. Supriya, C. T. Kamala, N. S. Chary, T. N. Rao, and Y. Anjaneyulu, "Speciation, accumulation of heavy metals in vegetation grown on sludge amended soils and their transfer to human food chain—a case study," Toxicological and Environmental Chemistry, vol. 82, no. 1-2, pp. 33–43, 2001.
4. Z. L. He, M. Zhang, X. E. Yang, and P. J. Stoffella, "Release behavior of copper and zinc from sandy soils," Soil Science Society of America Journal, vol. 70, no. 5, pp. 1699–1707, 2006. View at Publisher ·View at Google Scholar · View at Scopus
5. R. G. McLaren and L. M. Clucas, "Fractionation of copper, nickel, and zinc in metal-spiked sewage sludge," Journal of Environmental Quality, vol. 30, no. 6, pp. 1968–1975, 2001.
6. A. K. Gupta, S. Sinha, A. Basant, and K. P. Singh, "Multivariate analysis of selected metals in agricultural soil receiving UASB treated tannery effluent at Jajmau, Kanpur (India)," Bulletin of Environmental Contamination and Toxicology, vol. 79, no. 5, pp. 577–582, 2007.
7. A. Qishlaqi and F. Moore, "Statistical analysis of accumulation and sources of heavy metals occurrence in agricultural soils of Khoshk river banks, Shiraz, Iran," American-Eurasian Journal of Agricultural & Environmental Sciences, vol. 2, pp. 565–573, 2007.
8. G. Shen, Y. Lu, M. Wang, and Y. Sun, "Status and fuzzy comprehensive assessment of combined heavy metal and organo-chlorine pesticide pollution in the Taihu Lake region of China," Journal of Environmental Management, vol. 76, no. 4, pp. 355–362, 2005.
9. Q. W. Yang, C. Y. Lan, H. B. Wang, P. Zhuang, and W. S. Shu, "Cadmium in soil-rice system and health risk associated with the use of untreated mining wastewater for irrigation in Lechang, China,"Agricultural Water Management, vol. 84, no. 1-2, pp. 147–152, 2006.
10. Q. W. Yang, C. Y. Lan, and W. S. Shu, "Copper and Zinc in a paddy field and their potential ecological impacts affected by wastewater from a lead/zinc mine, P. R. China," Environmental Monitoring and Assessment, vol. 147, no. 1–3, pp. 65–73, 2008.
11. "International Rice Research Institute Las Banas IRRI," Institute Las Banas, Phillipines, 2005.

12. N. Prakongkep, A. Suddhiprakarn, I. Kheoruenromne, and R. J. Gilkes, "Micromorphological properties of Thai paddy soils," Kasetsart Journal : Natural Science, vol. 41, no. 5, pp. 42–48, 2007.
13. M. Zazoli and A. Edris, "Determination and estimation of Cd intake from Tarom rice," Journal of Applied Sciences and Environmental Management, vol. 10, pp. 147–150, 2006.
14. M. Radojevic andV. N. Bashkin, Practical Environmental Analysis, Royal Society of Chemistry, Cambridge UK, 1999.
15. M. Chen, X. M. Li, Q. Yang et al., "Total concentrations and speciation of heavy metals in municipal sludge from Changsha, Zhuzhou and Xiangtan in middle-south region of China," Journal of Hazardous Materials, vol. 160, no. 2-3, pp. 324–329, 2008.
16. Statsoft inc, "STATISTICA software for windows," Computer Program Manual, Tulsa, Okla, USA, 1999.
17. S. R. Tariq, M. H. Shah, N. Shaheen, A. Khalique, S. Manzoor, and M. Jaffar, "Multivariate analysis of selected metals in tannery effluents and related soil," Journal of Hazardous Materials, vol. 122, no. 1-2, pp. 17–22, 2005.
18. C. Kabala and B. R. Singh, "Fractionation and mobility of copper, lead, and zinc in soil profiles in the vicinity of a copper smelter," Journal of Environmental Quality, vol. 30, no. 2, pp. 485–492, 2001.
19. H. Horiguchi, E. Oguma, S. Sasaki et al., "Dietary exposure to cadmium at close to the current provisional tolerable weekly intake does not affect renal function among female Japanese farmers,"Environmental Research, vol. 95, no. 1, pp. 20–31, 2004.
20. G. Petruzzelli, "Recycling wastes in agriculture: heavy metal bioavailability," Agriculture, Ecosystems and Environment, vol. 27, no. 1–4, pp. 493–503, 1989.
21. NEQS, "Pakistan environmental protection agency," Government of Pakistan, 1999.
22. E. Gimeno-García, V. Andreu, and R. Boluda, "Heavy metals incidence in the application of inorganic fertilizers and pesticides to rice farming soils," Environmental Pollution, vol. 92, no. 1, pp. 19–25, 1996.

23. "Council of the European Communities (CEC)," Directive N 278 of 6/12/1986, Commission of the European Communities, Brussels, Belgium, 1986.
24. J. Hart, R. Fletcher, C. Landgren, D. Horneck, S. Webster, and M. Bondi, Christmas Tree Nutrient Management Guide For Western Oregon and Washington, EM 8856-E, Oregon State University Extension, 2004.
25. R. S. Oremland and J. F. Stolz, "The ecology of arsenic," Science, vol. 300, no. 5621, pp. 939–944, 2003.
26. S. Silver and L. T. Phung, "Genes and enzymes involved in bacterial oxidation and reduction of inorganic arsenic," Applied and Environmental Microbiology, vol. 71, no. 2, pp. 599–608, 2005.
27. P. M. Fox and H. E. Doner, "Trace element retention and release on minerals and soil in a constructed wetland," Journal of Environmental Quality, vol. 31, no. 1, pp. 331–338, 2002.
28. P. L. Smedley and D. G. Kinniburgh, "A review of the source, behaviour and distribution of arsenic in natural waters," Applied Geochemistry, vol. 17, no. 5, pp. 517–568, 2002.
29. B. Casentini, S. J. Hug, and N. P. Nikolaidis, "Arsenic accumulation in irrigated agricultural soils in Northern Greece," Science of the Total Environment, vol. 409, pp. 4802–4810, 2011.
30. S. Bhattacharya, K. Gupta, S. Debnath, U. C. Ghosh, D. Chattopadhyay, and A. Mukhopadhyay, "Arsenic bioaccumulation in rice and edible plants and subsequent transmission through food chain in Bengal basin: a review of the perspectives for environmental health," Toxicological and Environmental Chemistry, vol. 94, no. 3, pp. 429–441, 2012.
31. S. K. Acharyya, S. Lahiri, B. C. Raymahashay, and A. Bhowmik, "Arsenic toxicity of groundwater in parts of the Bengal basin in India and Bangladesh: the role of Quaternary stratigraphy and Holocene sea-level fluctuation," Environmental Geology, vol. 39, no. 10, pp. 1127–1137, 2000.
32. T. R. Chowdhury, G. K. Basu, B. K. Mandal et al., "Arsenic poisoning of Bangladesh groundwater,"Nature, vol. 401, pp. 545–547, 1999.

33. H. M. Anawar, J. Akai, T. Yoshioka et al., "Mobilization of arsenic in groundwater of Bangladesh: evidence from an incubation study," Environmental Geochemistry and Health, vol. 28, no. 6, pp. 553–565, 2006.
34. W. B. Achiba, N. Gabteni, A. Lakhdar et al., "Effects of 5-year application of municipal solid waste compost on the distribution and mobility of heavy metals in a Tunisian calcareous soil," Agriculture, Ecosystems and Environment, vol. 130, no. 3-4, pp. 156–163, 2009.
35. S. S. Huang, Q. L. Liao, M. Hua et al., "Survey of heavy metal pollution and assessment of agricultural soil in Yangzhong district, Jiangsu Province, China," Chemosphere, vol. 67, no. 11, pp. 2148–2155, 2007.
36. P. Zhuang, M. B. McBride, H. Xia, N. Li, and Z. Li, "Health risk from heavy metals via consumption of food crops in the vicinity of Dabaoshan mine, South China," Science of the Total Environment, vol. 407, no. 5, pp. 1551–1561, 2009.
37. J. Liu, K. Li, J. Xu et al., "Lead toxicity, uptake, and translocation in different rice cultivars," Plant Science, vol. 165, no. 4, pp. 793–802, 2003.
38. I. A. Katsoyiannis and A. A. Katsoyiannis, "Arsenic and other metal contamination of groundwaters in the industrial area of thessaloniki, Northern Greece," Environmental Monitoring and Assessment, vol. 123, no. 1-3, pp. 393–406, 2006.
39. M. R. Beychok, "A data base of dioxin and furan emissions from municipal refuse incinerators,"Atmospheric Environment. Part A. vol. 21, no. 1, pp. 29–36, 1987.
40. "US EPA estimating risk from contaminants contained in agricultural fertilizers," Draft Report, Office of Solid Waste, US Environmental Protection Agency, 1999, http://www.epa.gov/epawaste/index.htm.
41. G. C. Sigua, M. B. Adjei, and J. E. Rechcigl, "Cumulative and residual effects of repeated sewage sludge applications: forage productivity and soil quality implications in South Florida, USA," Environmental Science and Pollution Research, vol. 12, no. 2, pp. 80–88, 2005.

42. J. W. C. Wong, K. L. Li, L. X. Zhou, and A. Selvam, "The sorption of Cd and Zn by different soils in the presence of dissolved organic matter from sludge," Geoderma, vol. 137, no. 3-4, pp. 310–317, 2007.
43. S. Y. Selivanovskaya and V. Z. Latypova, "The use of bioassays for evaluating the toxicity of sewage sludge and sewage sludge-amended soil," Journal of Soils and Sediments, vol. 3, no. 2, pp. 85–92, 2003.
44. K. Chander and P. C. Brookes, "Effects of heavy metals from past applications of sewage sludge on microbial biomass and organic matter accumulation in a sandy loam and silty loam U.K. soil," Soil Biology and Biochemistry, vol. 23, no. 10, pp. 927–932, 1991.
45. X. Liu, S. Zhang, W. Wu, and H. Liu, "Metal sorption on soils as affected by the dissolved organic matter in sewage sludge and the relative calculation of sewage sludge application," Journal of Hazardous Materials, vol. 149, no. 2, pp. 399–407, 2007.
46. S. Bozkurt, L. Moreno, and I. Neretnieks, "Long-term fate of organics in waste deposits and its effect on metal release," Science of the Total Environment, vol. 228, no. 2-3, pp. 135–152, 1999
47. M. B. McBride, B. K. Richards, T. Steenhuis, and G. Spiers, "Long-term leaching of trace elements in a heavily sludge-amended silly clay loam soil," Soil Science, vol. 164, no. 9, pp. 613–623, 1999
48. A. Bhogal, F. A. Nicholson, B. J. Chambers, and M. A. Shepherd, "Effects of past sewage sludge additions on heavy metal availability in light textured soils: implications for crop yields and metal uptakes," Environmental Pollution, vol. 121, no. 3, pp. 413–423, 2003.
49. M. J. Sánchez-Martín, M. García-Delgado, L. F. Lorenzo, M. S. Rodríguez-Cruz, and M. Arienzo, "Heavy metals in sewage sludge amended soils determined by sequential extractions as a function of incubation time of soils," Geoderma, vol. 142, no. 3-4, pp. 262–273, 2007.

Chapter 6

Soil Fertility, Phosphorus Fractions, and Maize Yield as Affected by Poultry Manure and Single Superphosphate

A. O. Ojo[1], M. T. Adetunji[2], K. A. Okeleye[2], and C. O. Adejuyigbe[2]

[1]Institute of Agricultural Research and Training, Obafemi Awolowo University, P.M.B 5029, Moor Plantation, Ibadan, Oyo State, Nigeria

[2]Federal University of Agriculture, Abeokuta, Nigeria

ABSTRACT

A field experiment was conducted in 2007 and 2008 on a slightly acidic alfisol. Poultry manure (PM) was applied at 0, 5 t ha^{-1}, 10 t ha^{-1}, 15 t ha^{-1}, and 20 t ha^{-1} in combination with SSP at 0, 15 kg P ha^{-1}, 30 kg P ha^{-1}, 45 kg P ha^{-1}, and 60 kg P ha^{-1}, which was replicated three times. The pH and organic C were significantly increased by the application of PM alone while available P was highly increased by the sole application of SSP. Plant tissue P was significantly increased with the application of 30 kg P ha^{-1} while the largest grain yield was obtained when PM at 20 t ha^{-1} was combined with SSP at 60 kg P ha^{-1}. The buildup of organic P was observed when PM was applied at 15 t ha^{-1} while the combination of the two treatments increased residual P and Fe-P. However, P occlusion was effectively reduced with the sole application of PM. Organic P and residual P however had a strong positive relationship with the grain yield. Comparing the sole and combined application of the treatments, the combined application was more effective for most of the parameters observed.

INTRODUCTION

Phosphorus (P) is one of the essential elements for plant growth and the number of plants available P in the soil is often inadequate to meet plant requirements. Phosphorus is a macronutrient that plays a number of important roles in plant. Adequate phosphorus results in higher grain production, improved crop quality, greater stalk strength, increased root growth, and earlier crop maturity. Mokwunye et al. [1] and Warren [2] stated that phosphorus deficiency is one of the largest constraints to food production in tropical African soils due to low native P and high fixation by iron and aluminum oxides. It has also been established that phosphorus is relatively unavailable for plant uptake in highly weathered soils like the Ultisols [3]. In order to meet the need of phosphorus in these soils, farmers make use of inorganic fertilizers for their crops. However, P fertilizer is not readily available to these farmers because of scarcity in some cases and also due to the high cost of these fertilizers. Recently, farmers especially in the developed countries make use of organic manure which has been found to be effective like inorganic fertilizers for the release of some nutrients especially phosphorus but the resultant yield has been found not to be as high as when inorganic fertilizer is combined with the organic manure. Combining organic manure with P fertilizers has been found to not only increase yield but also reduce P fixation and subsequently increase phosphorus availability and uptake [3]. The study however aims at studying the effect of combined application of poultry manure and single superphosphate on phosphorus forms, soil fertility, and the yield maize.

MATERIALS AND METHODS

The field experiment was conducted at I.A.R&T experimental field at Ibadan in 2007 and 2008 early planting season (see Table 1). The experimental design was a 5 × 5 factorial experiment in a RCBD replicated three times. The plot size was 5 m × 4 m (20 m^2) with spacing of 75 cm × 50 cm to give a total plant population of 106,666 plants ha^{-1}. Three seeds of maize (TZB-SR) were planted after 2 weeks of treatment incorporation. The seedlings were thinned to two per stand two weeks after planting. The plots were not tilled in the early season (May) for the second year (2008) in order to avoid mixing up of treatments. Gramoxone and primextra were used as herbicides. Weeding was done at the 4th and 8th week after planting.

Table 1: Source of the soil used

Location	State	Coordinates	Vegetation	Soil series	Soil class (USDA)
Ibadan	Oyo	Lat. 7°30 Long. 3°54	Rainforest	Iwo	Ferrudulf

Leaf sampling of maize was done at eight weeks after planting. Leaf just below and opposite the ear was sampled randomly per plot. Ears of maize were allowed to dry on the stalk. Ears in the four middle rows of each plot were harvested (making allowance for two guard rows) on each side of the plot. The cobs were dehusked and weighed. Ten cobs per plots were randomly selected for cob weight measurement. The cobs were sun dried and the grains were weighed. Yield per hectare was computed at 12% moisture content for maize. Soil samples were analyzed before and after for phosphorus fractions and soil nutrient composition.

Soil Physical and Chemical Analysis

Particle size distribution was determined by the hydrometer method [4] using sodium hexametaphosphate as the dispersing agent. Soil pH was determined in distilled water (1 : 1 soil water ratio) with a pH meter. Exchangeable cations (K, Na, Ca, and Mg) by extraction, with ammonium acetate (pH = 7). Available P was extracted with 0.03 N NH_4F in 0.025 N HCl solution [5] and P in the extract was analysed colorimetrically by the molybdenum blue method at 660 nm. Soil organic carbon was determined by wet oxidation with sulphuric acid [6]. Total P was determined by wet digestion using perchloric acid and nitric acid while organic P was by ignition method. Phosphorus concentration in the digests was determined using the ascorbic acid method.

Phosphorus Fractionation Study

Fractionation procedure by Chang and Jackson [7] was used to quantify different forms of inorganic P in the studied soils (Table 2). Fractionation was done on the soil before and after amendment for incubation, greenhouse, and field experiment. The sequence of the chemical extraction and a brief description of the targeted P forms are shown in Table 2.

Table 2: Sequential P fractionation procedures and targeted P forms

Location	Extract ants	Equilibration	Washing	Targeted
NH4Cl-P	IM NH4Cl	30 mins	None	Saloid-bound P (loosely bound P)
NH4F-P	0.5 M NH4F	1 hour	None	Al-P
Na OH-P	0.1 M NaOH saturated Nacl + 5 drops of conc. H2SO4	15 mins	Saturated Nacl	Fe-P

Sodium + citrate Sodium dithionite P	0.3 M sodium citrate + 1 g solid Sodium dithionate + sat Nacl	Shake 15 mins, preheat 15 mins at 85°C after sodium citrate, additional 15 mins after dithionite addition	Saturated Nacl	Reductant-soluble P
NaOH-P	0.1 M NaOH + few drops of conc. H2SO4 to remove colour	15 mins	Saturated NaCl	Occluded P
H2SO4-P	0.25 M H2SO4 + 25 mL Sat. Nacl	1 hour	Saturated Nacl	Ca-P
Residual P	Conc. HNO3 + HCl + 30% H2O2	Variable until complete	None	Organically stable organic and inorganic P

Determination of Phosphorus in the Extracts

The phosphorus in the extracts was determined by the methyl blue colour method of Murphy and Riley [8]. An aliquot of 10 mL was used in developing the colour. The determination of P in NH_4F, NaOH, and DCB extracts however requires the following pretreatment before the development of the colour.

- NH_4F extract: 0.05 g of H_3BO_3 per 50 mL of the extract was added before taking the aliquot.
- NaOH extract: after precipitating the organic matter with conc. H_2SO_4, the pH of the extract was adjusted with NaOH using 2,4-dinitrophenol indicator. An aliquot of the clear solution was then taken into a 50 mL volumetric flask.

- extract (reductant soluble P): 10 mL of the extract was pipetted into a test tube and 1 mL of $HClO_4$ was added to it and boiled to destroy the dithionite in the extract. It was then transferred into a 50 mL volumetric flask and 3 mL of 5% ammonium molybdate solution was added while distilled water was used to make up to mark. Absorbance was read after 30 minutes.

Analysis of Poultry Manure Sample

Total analysis of the sample was done by weighing 0.5 g of poultry manure, adding 10 mL of $HNO_3/HClO_4$, (2 : 1), and digesting at 150°C. The temperature was increased at 1.5 hours to 230°C and 2 mL of HCl/H_2O (1 : 1) was added. The digestion was now continued for another 30 minutes [9].

Statistical Analysis

The data collected were subjected to analysis of variance using the statistical analysis system (SAS)—general linear model [10]. Means were separated by Duncan multiple range test.

RESULTS

Initial Chemical Properties of the Soil Used

The pH of the soil was slightly acidic and therefore a need for the application of organic manure. The soil was low in N, Na, and organic carbon while P was moderately available in it. It was however adequate with cations like Ca, Mg, and K (Table 3).

Table 3: Chemical and physical properties of the soil used for the study

Parameters	
pH (H2O)	6.4
O.M (g kg–1)	24.08

N (g kg–1)	0.2
P (mg kg–1)	14.2
Exchangeable cations (c molc kg–1)	
Ca	1.14
Mg	0.82
K	0.47
Na	0.09
Particle size (g kg–1)	
Sand	740
Clay	120
Silt	140

Characterization of the Poultry Manure Used and the Initial Phosphorus Fractions of the Soil

Nitrogen in the poultry manure used was 5.82% which was similar to the result observed for phosphorus (5.54%) while potassium was low in the soil (0.94%). Organic C had a high percentage of 13.38% which led to the resultant C/N ratio of 2.3 (Table 4). Also notable was the result obtained for Fe (1555 mg/kg).

Table 4: Characterization of the poultry manure used

pH	7.03
Nitrogen (N) (%)	5.82
Phosphorus (P) (%)	5.54
Potassium (K) (%)	0.94
Na (%)	1.3
SO4-S (%)	0.14
Organic C (%)	13.34
C/N	2.3
Calcium (%)	8.04
Magnesium (%)	0.61
Iron (mg kg–1)	1555

Copper (mg kg–1)	33.3
Zinc (mg kg–1)	100.3
Manganese (mg kg–1)	180

The order of abundance of the phosphorus fractions before the application of treatments was organic P > Fe-P ≡ occluded P > residual P (Table 5).

Table 5: Phosphorus fractions of the soil before planting (mg kg^{-1})

Phosphorus fractions (mg kg–1)	
Total P	272.84
Organic P	45.90
Fe-P	33.39
Occluded P	33.39
Residual P	16.16

Effect of Combined Application of Poultry Manure and SSP on the Soil Chemical Properties

There was a decrease in pH as a result of the sole application of SSP in 2007, which further decreased in 2008 (Table 6). Application of poultry manure alone increased the pH of the soil significantly when compared to the other treatments. However, the combined application of poultry manure and SSP increased the pH of the soil in 2007 with a further increase in 2008. In other words, poultry manure alone can be compared favorably with the combined application of poultry manure and SSP in increasing soil pH especially in slightly acidic soil. Effect of SSP on organic C was not significant and decreased when compared to the initial value. However, the application of poultry manure increased organic C significantly when poultry manure was applied alone in 2007. In 2008, organic C further increased with the sole application of poultry manure. The effect of the combined application of poultry manure and SSP was not significant on organic carbon in 2007 but significant in 2008 with the combined application of poultry manure and SSP. Available P increased over the control either with the sole application of poultry manure or SSP or when the two treatments were combined. However, in 2007 the sole application of SSP was more effective in increasing available P than the sole application of poultry manure but was the most effective with the combined application of poultry manure and SSP, while, in 2008, the combined treatments were not significant but the sole application was.

Table 6: Effect of poultry manure (PM) and SSP on some soil chemical properties in 2007 and 2008

Treatments PM (tha−1)	Treatments SSP (kgPha−1)	pH (2007)	pH (2008)	Organic C, % (2007)	Organic C, % (2008)	Available P, mgkg−1 (2007)	Available P, mgkg−1 (2008)
0	0	6.23	6.30	0.19	0.16	5.90	6.07
	15	6.73	6.73	0.62	0.85	92.57	96.78
	30	6.32	6.30	1.02	1.27	39.41	39.91
	45	6.51	6.51	1.26	0.74	69.60	70.02
	60	6.42	6.87	0.75	0.89	42.56	41.18
5	0	7.06	7.06	0.65	1.05	41.50	41.70
	15	6.53	6.53	0.92	1.90	56.39	56.27
	30	6.52	6.52	1.06	1.42	43.56	43.84
	45	6.65	6.62	1.12	1.29	90.05	87.66
	60	6.92	6.92	0.91	0.97	64.69	65.82
10	0	6.87	6.87	1.03	1.39	64.82	63.30
	15	6.82	6.73	1.03	1.04	59.62	59.74
	30	6.72	6 72	1.37	0.87	67.06	66.74
	45	6.86	6.86	0.98	1.34	83.32	81.65
	60	6.87	6.87	1.02	1.53	60.42	60.42
15	0	6.50	6.50	1.63	1.52	40.05	96.78
	15	6.87	6.93	1.66	2.65	67.71	67.78
	30	6.80	6.80	1.08	1.45	96.82	87.49
	45	6.65	6.72	1.08	1.62	88.31	87.24
	60	6.62	7.16	1.25	1.50	75.57	74.54
20	0	6.93	6.93	1.03	1.75	59.60	53.07
	15	6.60	6.60	0.78	1.01	72.68	76.81
	30	6.60	6.60	1.22	1.26	72.68	79.69
	45	6.50	6.50	0.82	1.12	64.87	64.44
	60	6.97	6.96	1.23	2.60	61.74	94.84

	LSD ($P \leq$ 0.05)						
	SSP	*	*	NS	*	*	NS
	PM	*	*	*	*	*	NS
	SSP × PM	*	NS	NS	*	*	NS

*Significant at $P \leq 0.05$, NS: not significant.

Effect of the Combined Application of Poultry Manure and SSP on Plant Tissue P Concentration and Grain Yield of Maize

Effect of poultry manure and SSP alone with their combinations was significant in both years (Table 7). In 2007, plant tissue P was high when the treatments were applied solely but combined application was highly significant. Plant tissue P continued to increase in 2008 with the sole and combined application but with the most significant increase with the sole application of SSP.

Table 7: Effect of poultry manure (PM) and SSP on plant tissue concentration and grain yield in 2007 and 2008

Treatments		Plant tissue (%) P		Grain yield (t ha−1)	
PM (t ha−1)	SSP (kg P ha−1)	2007	2008	2007	2008
0	0	0.43	0.53	1.33	0.81
	15	0.94	1.04	3.12	3.21
	30	0.82	2.51	2.29	1.45
	45	1.29	2.02	2.57	1.05
	60	0.79	1.93	2.39	1.42
5	0	0.88	0.77	2.76	1.98
	15	1.97	1.56	2.61	1.25
	30	0.87	0.96	2.69	1.67
	45	0.85	2.50	2.71	1.74
	60	0.82	1.24	2.35	2.57

10	0	1.32	0.86	2.87	1.66
	15	1.49	1.82	2.46	2.18
	30	1.01	1.33	2.59	3.02
	45	1.61	0.84	2.48	1.78
	60	1.28	1.35	2.42	3.53
15	0	2.00	1.04	3.63	0.83
	15	1.82	1.01	2.75	3.85
	30	0.87	0.83	2.75	3.50
	45	1.49	1.42	2.60	3.21
	60	0.93	1.02	2.20	3.76
20	0	1.03	0.94	2.82	2.82
	15	0.87	1.36	2.86	2.41
	30	1.36	1.52	2.64	2.44
	45	0.99	2.09	2.36	2.30
	60	0.99	1.35	2.37	3.99
	LSD ($P \leq 0.05$)				
	SSP	*	*	NS	*
	PM	*	*	*	*
	SSP × PM	*	*	*	NS

*Significant at $P \leq 0.05$, NS: not significant.

The sole application of SSP had no significant effect on the grain yield of maize in 2007 while the sole application of poultry manure and the combined application of the two treatments were significant in the same year. However, in 2007, the highest grain yield of maize was obtained by the sole application of poultry manure. In 2008, effect of the sole application of poultry manure and SSP on maize grain yield was significant while the combined application was not. The combined application of the two treatments gave the highest grain yield in 2008.

Effect of the Combined Application of Poultry Manure and SSP on Phosphorus Fractions

Effect of the sole and combined application of poultry manure and SSP was significant in both 2007 and 2008 (Table 8). In 2007, organic P was significantly increased over the initial value when poultry manure was applied solely than when SSP was applied alone. Combining the two treatments also increased organic P significantly in 2007. The application of poultry manure alone however gave the highest increase in organic P in 2007. In 2008, organic P reduced with the sole application of SSP but increased when poultry manure was applied alone and with some combined treatments. The sole application of poultry manure was again effective in 2008 in increasing organic P. Residual P increased over the control when poultry manure and SSP were applied alone and when combined in both years. Residual P increased in 2007 with the combined application of the two treatments but reduced in 2008. A similar trend to that of residual P was observed for Fe-P but with few exceptions when the 2 treatments were combined. In 2008, the largest buildup of Fe-P was observed with the combined application of poultry manure and SSP. In both years of cropping, sole application of poultry manure was effective in reducing P occlusion more than the combined application.

Table 8: Effects of poultry manure and SSP on phosphorus fractions in 2007 and 2008

Treatments		Phosphorus fractions (mg kg−1)							
PM (t ha−1)	P rates (Kg P ha−1)	Organic P 2007	Organic P 2008	Residual P 2007	Residual P 2008	Fe-P 2007	Fe-P 2008	Occluded P 2007	Occluded P 2008
0	0	49.91	50.04	21.83	12.26	38.50	19.48	27.25	28.03
	15	128.96	112.50	36.41	18.55	88.21	21.62	31.61	10.11
	30	135.63	125.02	36.08	17.54	90.24	17.74	22.66	10.53
	45	115.63	140.63	40.73	20.52	68.09	64.92	31.65	11.72
	60	177.50	130.58	54.99	22.97	117.37	18.86	36.72	12.24
5	0	75.03	111.27	47.43	17.45	88.12	50.76	16.61	9.36
	15	138.75	168.75	37.18	15.52	86.82	16.72	19.79	11.08
	30	167.35	146.85	41.47	17.34	88.44	16.85	26.08	9.56
	45	143.14	116.22	48.76	19.31	87.08	137.76	27.82	12.34
	60	100.06	106.83	46.26	19.22	116.29	62.25	21.18	17.14
10	0	105.07	160.05	49.32	17.43	87.15	86.81	18.63	10.52
	15	180.05	118.75	45.24	16.45	88.31	19.07	26.82	8.51
	30	183.73	155.05	51.29	18.63	115.23	15.35	26.18	10.33
	45	160.37	122.50	38.16	21.28	104.62	117.55	44.02	12.88
	60	145.61	130.11	42.28	20.75	117.69	148.94	30.12	11.72

15	0	198.75	112.50	36.61	16.92	61.81	21.62	12.14	17.01
	15	91.85	120.02	45.57	20.07	97.69	18.25	22.89	8.51
	30	124.38	159.38	41.89	18.62	117.68	15.96	38.42	10.64
	45	105.53	174.33	40.11	25.54	98.54	131.93	42.11	14.26
	60	161.86	170.52	34.45	22.76	202.36	152.11	28.95	9.92
20	0	149.35	103.77	47.51	18.91	84.52	16.08	12.46	12.13
	15	99.39	101.88	44.41	16.07	75.13	18.95	32.88	9.37
	30	141.85	174.83	42.39	21.05	117.99	15.97	34.91	11.08
	45	195.00	136.22	54.63	31.62	89.62	13.18	36.81	11.21
	60	171.25	150.02	57.24	23.65	113.12	125.41	39.37	13.21
	LSD ($P \leq 0.05$)								
	SSP	*	*	*	*	*	*	*	*
	PM	*	*	*	*	*	*	*	*
	SSP × PM	*	*	*	*	*	*	*	*

*Significant at $P \leq 0.05$, NS: not significant

Relationship among the P Fractions, Soil Chemical Properties, and Maize Grain Yield

In 2007, the pH of the soil had a strong negative relationship with Al-P and occluded P but positive relationship with residual P (Table 9), while in 2008, increase in the pH of the soil favored the buildup of Fe-P (Table 10). Available P in 2007 had a positive correlation with organic P, Ca-P, and Fe-P but a negative correlation with occluded P. Organic C had no significant relationship with the P fractions in 2007 but a positive correlation with organic P and a negative correlation with occluded P in 2008. Organic P and residual P contributed positively to the maize grain yield in both 2007 and 2008.

Table 9: Correlation coefficient (r) among P fractions, soil chemical properties, and maize grain yield in 2007

Soil properties	P fractions (mg kg−1)			
	Organic P	Fe-P	Occluded P	Residual P
pH	0.098	0.201	−0.550	0.633
Available P	0.514	0.433	−0.528	−0.071
Organic C	0.206	0.115	−0.031	−0.105
Grain yield	0.413	0.145	0.034	0.402

Table 10: Correlation coefficient (r) among P fractions, soil chemical properties, and maize grain yield in 2008

Soil properties	P fractions (mg kg−1)			
	Organic P	Fe-P	Occluded P	Residual P
pH	0.118	0.458	0.065	−0.144
Available P	0.053	0.173	−0.753	0.048
Organic C	0.500	0.122	−0.557	−0.177
Grain yield	0.403	0.105	0.123	0.412

DISCUSSION

Initial Chemical Properties of the Soil Used

The pH of the soil was slightly acidic and this will make Fe-P be more available than the other inorganic P fractions. Although phosphorus was moderately available, poultry manure and SSP were further applied in order to study the effect of continuous application of organic and inorganic fertilizer on such soils yearly as has been practiced by many farmers in Nigeria.

Characterization of the Poultry Manure Used and Initial Phosphorus Fractions of the Soil

A common and traditional practice has utilized manure as an organic fertilizer or soil improver due to its contribution of organic matter and nutrients, mainly N and P [11]. The high percentage of nitrogen in the poultry manure used could have been due to the decomposition of poultry manure during the curing stage before application to the soil with a subsequent release of N which was detected during the analysis. Manure in past studies has been applied according to the nitrogen needs of the soil [12]. Macronutrients such as phosphorus are part of the advantage of using organic waste in agriculture, which may be recovered and incorporated into natural cycle and this was evident in the percentage of phosphorus in the poultry manure used. It also contains a high level of organic matter which generates changes in chemical, physical, and biological soil amendment properties [12].

Organic P was the highest P fraction at the initial stage of this study. About two-thirds of the phosphorus in fresh manure is in organic form [12]. The organic P pool is usually considered a major source of organic phosphorus and can be converted to inorganic P by ignition or wet digestion. Initial abundance of Fe-P could be attributed to the acidic nature of the soil; however occluded P also had the same value which could mean there is an equilibrium reaction between the labile and the nonlabile P fractions.

Effect of Combined Application of Poultry Manure and SSP on the Soil Chemical Properties

Acidity in sols if not properly managed will eventually make the soil be low or not yielding and therefore a need for amendment. In this study, application of poultry manure compared favourably with the combined application of poultry manure and SSP in increasing the pH of the soil while applying inorganic fertilizer alone decreased the soil pH. Soil pH of an acidic Ultisol in the eastern Nigeria was observed to increase soil pH [13]. Sole and combined organic and inorganic fertilizer has been found to increase soil nutrients favorably. The combined application of the two treatments was highly effective in increasing soil organic C in both 2007 and 2008. Earlier studies on poultry manure ability to increase soil organic C have been reported [13, 14], likewise when it is combined with SSP [15]. Single superphosphate was however more effective than the sole application of poultry manure for increasing the available P content of the soil although the combined application of the two treatments still gave the highest value [16]. The effectiveness of SSP for available P over poultry manure could be due to the immediate release of phosphate into the soil when SSP is being applied.

Effect of the Combined Application of Poultry Manure and SSP on Plant Tissue P Concentration and Grain Yield of Maize

High plant tissue P was obtained in 2007 and 2008 with the sole and combined application of poultry manure and SSP. However, the sole application of SSP was however sustainable in the 2nd year to cause a further increase in the concentration of P in the shoot. A similar result has been observed by Adeniyan and Ojeniyi, [17]. The increase in the plant tissue P in 2008 was as a result of P accumulation observed in both years, which would have resulted in the transfer to the shoot of the plant instead of being washed into nearby river and hereby preventing eutrophication.

When the treatments were applied solely, application of poultry manure alone was more effective than SSP, that is, considering the high grain yield. Although this was in the first year of planting, the combined application of poultry manure and SSP was again sustainable in the second year. The combined application of organic and inorganic resources has been found to favor maize production [14].

Effect of the Combined Application of Poultry Manure and SSP on Phosphorus Fractions

The buildup of organic P was observed in both years of this study with the sole application of poultry manure and this signifies the importance of organic manure when considering amendments for an acidic soil. Addition of soybean residue and wheat residue or in combination with fertilizer P favored the buildup of organic P in a study [18]. The inorganic P fractions considered, that is, residual P and Fe-P, was significantly increased when poultry manure was combined with SSP. This indicates the ability of the combined application of poultry manure and SSP to maintain the fertility of the soil through increase in the labile P fraction [19]. Residual P however reduced in 2008 and this was also observed by Agbenin and Goladi, [20] in an experiment involving the addition of P fertilizers or in combination with cow manure. Significant was the ability of poultry manure alone to reduce P occlusion in both years, that is, occluded P, which further confirms an earlier study which states that addition of organic materials to soil decreases P fixation [3]. Inorganic and organic products are generated during the partial decomposition of organic waste and organic acids can be adsorbed into the soil surface thereby decreasing the potential P adsorption by blocking the formation of complexes [21].

Relationship among the P Fractions, Soil Chemical Properties, and Maize Grain Yield

Organic P and Fe-P contributed significantly to the increase observed in available P in 2007 while occluded P decreased with an increase in available P. This further confirms the importance of poultry manure while considering soil amelioration. In both years of cropping, organic P was sustainable in increasing grain yield but Fe-P was only effective in 2007. Residual P (a nonlabile P fraction) equally contributed to the increase observed in grain yield in both years. In other words, there would have been an equilibrium reaction between the nonlabile P (residual P) and labile P (Fe-P) fraction which would have subsequently resulted in an increase in available P.

CONCLUSIONS

Application of poultry manure alone was effective in increasing the pH, organic C, and available P content of the soil. However, application of SSP was highly significant in increasing the available P content of the soil and this was evident in the increase obtained for the plant tissue P in 2008, while, for grain yield, the most effective was the combined application of poultry manure and SSP. Increase in organic P fraction was as a result of the application of poultry manure while the combined application of poultry manure and SSP increased the Fe-P fraction. Poultry manure alone was also the most effective treatment for the reduction of P occlusion in the soil. Significant were also the importance of poultry manure and the combined application of poultry manure and SSP exhibited through the strong relationship that existed among available P, Fe-P, and organic P. Although poultry manure was compared favorably with the combined application of poultry manure and SSP, the combination of the treatments was more suitable for the sustainability of the fertility of the slightly acidic soil.

REFERENCES

1. A. U. Mokwunye, S. H. Chien, and E. Rhodes, "Phosphorus reaction with tropical African soils," inManagement of Nitrogen and Phosphorus Fertilizers in Sub-Saharan African, A. U. Mokwanye and P. L. C. Vick, Eds., pp. 253–281, Martins Nulhoff, Dordrecht, The Netherlands, 1986.
2. G. Warren, Fertilizer Phosphorus Sorption and Residual Value in Tropical African Soils, NRI Bulletin 37, Natural Resource Institute, Chatham, UK, 1992.
3. F. Iyamuremye, R. P. Dick, and J. Baham, "Organic amendments and phosphorus dynamics: I. Phosphorus chemistry and sorption," Soil Science, vol. 161, no. 7, pp. 426–435, 1996.

4. G. J. Bouyoucos, "Hydrometer method improved for making particle size analysis of soils," Agronomy Journal, vol. 54, no. 5, pp. 464–465, 1962.
5. R. H. Bray and L. T. Kurtz, "Determination of total, organic and available forms of phosphorus in soils," Soil Science, vol. 59, pp. 39–45, 1945.
6. A. Walkey and I. A. Black, "An examination of the Degtjareff method for determining organic carbon in soils: effect of variations in digestion conditions and of inorganic soil constituents," Soil Science, vol. 63, pp. 251–263, 1934.
7. S. C. Chang and M. L. Jackson, "Fraction of soil phosphates," Soil Science, vol. 83, pp. 133–144, 1957.
8. J. Murphy and J. P. Riley, "A modified single solution method for the determination of phosphate in natural waters," Analytica Chimica Acta, vol. 27, pp. 31–36, 1962.
9. E. J. Udo and J. A. Ogunwale, Manual of Soil, Plant and Water Analysis, Department of Agronomy, University of Ibadan, Ibadan, Nigeria, 1978.
10. SAS Institute, SAS for Linear Models. A Guide to the ANOVA and GLM Procedures, SAS Institute, Cary, NC, USA, 2000.
11. Z. He, T. S. Griffin, and C. W. Honeycutt, "Evaluation of soil phosphorus transformations by sequential fractionation and phosphatase hydrolysis," Soil Science, vol. 169, no. 7, pp. 515–527, 2004.
12. B. J. Wienhold, "Changes in soil attributes following low phosphorus swine slurry application to no-tillage sorghum," Soil Science Society of America Journal, vol. 69, no. 1, pp. 206–214, 2005.
13. A. O. Ano and J. A. Agwu, "Effects of animal manures on selected soil chemical properties," Nigeria Journal of Soil Science, vol. 15, pp. 14–19, 2005.
14. L. R. Bulluck III, M. Brosius, G. K. Evanylo, and J. B. Ristaino, "Organic and synthetic fertility amendments influence soil microbial, physical and chemical properties on organic and conventional farms," Applied Soil Ecology, vol. 19, no. 2, pp. 147–160, 2002.

15. M. C. Manna, A. Swarup, R. H. Wanjari, B. Mishra, and D. K. Shahi, "Long-term fertilization, manure and liming effects on soil organic matter and crop yields," Soil and Tillage Research, vol. 94, no. 2, pp. 397–409, 2007.
16. H. Zhang, B. Wang, and M. Xu, "Effects of inorganic fertilizer inputs on grain yields and soil properties in a long-term wheat-corn cropping system in South China," Communications in Soil Science and Plant Analysis, vol. 39, no. 11-12, pp. 1583–1599, 2008.
17. O. N. Adeniyan and S. O. Ojeniyi, "Effect of poultry manure, NPK 15-15-15 and combination of their reduced levels on growth and soil chemical properties," Nigeria Journal of Soil Science, vol. 15, pp. 34–41, 2005.
18. D. D. Reddy, S. A. Rao, and M. Singh, "Changes in P fractions and sorption in an Alfisol following crop residues application," Journal of Plant Nutrition and Soil Science, vol. 168, no. 2, pp. 241–247, 2005.
19. T. Q. Zhang and A. F. MacKenzie, "Changes of soil phosphorous fractions under long-term corn monoculture," Soil Science Society of America Journal, vol. 61, no. 2, pp. 485–493, 1997.
20. J. O. Agbenin and J. T. Goladi, "Dynamics of phosphorus fractions in a savanna Alfisol under continuous cultivation," Soil Use and Management, vol. 14, no. 2, pp. 59–64, 1998.
21. M. S. Mkhabela and P. R. Warman, "The influence of municipal solid waste compost on yield, soil phosphorus availability and uptake by two vegetable crops grown in a Pugwash sandy loam soil in Nova Scotia," Agriculture, Ecosystems and Environment, vol. 106, no. 1, pp. 57–67, 2005.

Chapter 7

Raw Agro-industrial Orange Peel Waste as a Low Cost Effective Inducer for Alkaline Polygalacturonase Production from Bacillus licheniformis SHG10

Amira M Embaby[1], Aliaa A Masoud[2], Heba S Marey[3], Nadia Z Shaban[2], and Tayssir M Ghonaim[2]

[1]Department of Biotechnology, Institute of Graduate Studies and Research, University of Alexandria, 163 Horreya Avenue, Chatby 21526, Egypt

[2]Faculty of Science, Department of Biochemistry, University of Alexandria, Alexandria, Egypt

[3]Department of Environmental Studies, Institute of Graduate Studies and Research, University of Alexandria, Alexandria, Egypt

ABSTRACT

The current study underlines biotechnological valorization of the accumulated and the non-efficiently utilized agro-industrial orange peel waste to produce polygalacturonase (PGase), an industrially important enzyme with augmented demands in enzymes markets, from *Bacillus licheniformis* SHG10. Sequential statistical optimization of PGase production was performed through one variable at a time (OVAT) approach, Plackett-Burman (PB) and response surface methodology (RSM). The impact of introduction of six raw agro-industrial wastes (orange, lemon, banana, pomegranate, artichoke peel wastes and wheat bran) and other synthetic carbon sources separately into the fermentation broth on PGase productivity was studied through OVAT approach. Orange peel waste as sole raw carbon source in basal medium proved to be the best PGase inducer. It promoted PGase productivity with relative specific activity of 166% comparable with the effect imposed by synthetic citrus pectin as a reference inducer. Three key determinants (orange peel waste, pH of the production medium and incubation temperature) had RSM optimal levels of 1.76% (w/v), 8.0 and 37.8°C, respectively along with maximal PGase level (2.69 μg galacturonic acid. min^{-1}. mg^{-1}) within 48 hrs. Moreover, SHG10 PGase exhibited activity over a wide range of pH (3-11) and an optimal activity at 50°C. Data greatly encourage pilot scale PGase production from *B. licheniformis* SHG10.

BACKGROUND

Citrus fruit is one of the commercial crops in the Egyptian market (Mohamed et al. 2010). Orange juice is one of the most consumed beverages today (Martin et al. 2010). Consequently, a high percentage of citrus fruit is used for manufacturing of juice and marmalade. Approximately 50-60% of citrus fruit is transformed into citrus peel waste (Wilkins et al. 2007). This results in accumulation of large quantities of citrus peel waste as a by-product in citrus-processing industry. Accumulated large quantities of the orange peel waste along with environmental considerations to avoid health hazards derived from unsatisfactory disposal methods addressed the indispensable need for finding alternative biotechnological solutions for waste valorization (Martín et al. 2013; Rivas et al. 2008). According to current environmental legislation, any waste could be considered as raw material as long as there is an option to develop method for its valorization (Möller et al.2001). High value products could be manufactured upon using orange peel waste as a potentially valuable low cost resource (Martin et al. 2010; Rivas et al. 2008; Balu et al. 2012). Orange peel waste was reported to contain 16.9% soluble sugars, 9.21% cellulose, 10.5% hemi-cellulose, and 42.5% pectin as the most important components (Rivas et al. 2008). A vast number of promising methods encountered in efficient utilization of orange peel waste has been described thoroughly in the literature. Among these methods is employing this waste in enzymes industries (Siles and Thompson 2010).

The term pectinolytic enzymes (pectinases) is the generic name of a family of enzymes involved in the process of pectin degradation. This complex bioprocess is achieved mainly by a set of pectinolytic enzymes catalyzed reactions (e.g., hydrolysis, trans-elimination and de-esterification) of the ester bond between the carboxyl and the methyl ester groups of pectin (Rehman et al.2012). Four types (polygalacturonases, pectin lyases, pectate lyases and pectin methyl esterases) of these enzymes classified based on their mode of action are involved under the generic name of this family (Alkorta et al. 1998; Hoondal et al. 2002; Kuhad et al. 2004). These enzymes have a vast number of industrial applications in food (e.g., juice clarification, refinement of vegetables fibers, extraction of vegetables oil, curing of coffee and cocoa beans) (Silva et al. 2002; Gummadi and Panda 2003; Demir et al. 2012; Quattara et al. 2008; Pedrolli et al. 2009). biopulping of papers (Sittidilokratna et al. 2007) and textiles (Basu et al. 2009). Among pectinolytic enzymes, polygalacturonases (PGase), are the enzymes of particular interest to industry. Endo- PGase (E.C. 3.2.1.67) and exo- PGase ((E.C. 3.2.1.82) catalyze the hydrolysis of internal and external -1,4 glycosidic bond linking –galacturonic acid residues in pectin, respectively producing shorter pectin molecular structures, decreasing the viscosity, increasing the yield of juices, and determining the crystalline structure of the final product (Souza et al. 2003).

The up to date review of literature reported a vast number of microorganisms as PGase producers mainly fungi such as *Aspergillus spp, Rhizopus stolonifer, Alternaria mali, Fusarium oxysporum,Neurospora crassa, Penicillium italicum* ACIM F-152 and a little bit form bacteria confined to*Agrobacterium tumefaciens, Bacteroides thetaiotamicron, Ralstonia solanacearum, Bacillus spp*and *Enterobacter aerogenes* NBO2 (Jayani et al. 2005, 2010; Darah et al. 2013). Microbial pectinase(s) particularly those of fungal origin account for 25% of the global food and industrial enzymes sales (Demir et al. 2012). Although fungi are considered to be potent pectinases producers, but the drawbacks included in the physicochemical properties of these enzymes greatly limit the utilization of these enzymes on a wide industrial scale (Soares et al. 1999).

The underlying reasons behind carrying out the current study could be outlined in the following annotations; a) increased demand for commercial PGase(s) in the enzymes markets worldwide, b) indispensable need for valorization of the accumulated and the non-efficiently utilized raw agro-industrial orange peel waste in a biotechnological manner and c) necessity for continuous searching for novel PGase(s) with new characteristics to overcome the shortcomings involved in PGase(s) of fungal origin that greatly limit their utilization on a wide industrial scale. In this context, the present study aims to address sequential statistical optimization of PGase production from *Bacillus licheniformis* SHG10 strain upon using the raw agro-industrial orange peel waste as a sole PGase inducer and a sole carbon source in a very low cost effective medium.

RESULTS

OVAT Results

The influence of different agro-industrial wastes (e.g., orange, lemon, pomegranate, banana, artichokes peel wastes and wheat bran) and synthetic carbon sources (e.g., citrus pectin, glucose, fructose, maltose, xylose, glycerol, sucrose, peptone, beef extract and tryptone) on PGase production by *B. licheniformis* SHG10 strain was studied. Table 1 revealed that orange peel waste at a concentration of 1% (w/v) was the best co-inducer sole carbon source. It enhanced the production of PGase with a relative specific activity of 166% when compared to the effect of citrus pectin (0.5% w/v). While, the substitution of orange peel waste with lemon peel waste (1% w/v) resulted in improvement in PGase level with relative specific activity of 133% of that induced by citrus pectin. However, addition of wheat bran [(0.5% (w/v) and 1% (w/v)] or banana peel waste (1% w/v) separately instead of citrus pectin achieved a level of PGase almost quite similar to that obtained upon using citrus pectin. Conversely, other carbon sources alternative to citrus pectin such as pomegranate and artichoke peel wastes, glucose, fructose, maltose, xylose, sucrose, glycerol, peptone, beef extract and tryptone at the concentrations mentioned in Table 1 exerted suppressive effects on the productivity of PGase by *B. licheniformis* SHG10.

Table 1: Agro-industrial wastes and some synthetic carbon sources as PGase inducers

Parameter	Relative specific activity (%)a,b
Agro-industrial wastes (1% w/v):	
Orange peel waste	166.80
Lemon peel waste	133.56
Banana peel waste	105.80
Artichoke peel waste	55.17
Pomegranate peel waste	53.59
Synthetic carbon sources:	
Citrus pectin (0.5% w/v)	100.00
Wheat bran (0.5% w/v)	102.00
Wheat bran (1% w/v)	92.00
Glucose (0.5% w/v)	41.82
Fructose (0.5% w/v)	64.10
Maltose (0.5% w/v)	33.01
Xylose (0.5% w/v)	59.04
Sucrose (0.5% w/v)	25.60
Glycerol (0.5% w/v)	27.59
Peptone (0.5% w/v)	33.42
Peptone (1% w/v)	50.20
Beef extract (0.5% w/v)	14.11
Beef extract (1% w/v)	14.11
Tryptone (0.5% w/v)	39.45

Tryptone (1% w/v)	28.72

[a]Specific activity (U/mg) was determined as µg galacturonic acid $.min^{-1}$. mg $protein^{-1}$.

[b]All relative specific activities of PGase were related to U/mg PGase obtained upon using citrus pectin – based basal medium as production medium.

Embaby et al.

Embaby et al. SpringerPlus 2014 3:327, doi:10.1186/2193-1801-3-327

Moreover, the effect of addition of some salts as supplementary substances to the production medium was tested. Data of Table 2 demonstrated that separate introduction of $NaNO_3$, KNO_3, $CaCl_2$, $FeSO_4$ and $MgSO_4$ salts to the control production medium (citrus pectin based- basal medium) each at a concentration of 0.2% (w/v) exhibited diverse degrees of enhancement in the level of PGase with relative specific activity of 145%, 163.8%, 198,9%, 228.57% and 237.7%, respectively when compared to that level obtained upon using control production medium. The addition of $(NH_4)_2SO_4$ at a final concentration of 0.2% (w/v) resulted in a level of PGase almost equivalent to that of the control production medium. On the other hand, a low level of PGase was detected in the production medium containing NH_4Cl at a concentration of 0.2% (w/v).

Table 2: Effect of some medium supplements in citrus pectin based- mineral medium on PGase productivity

Parameter	Relative specific activity (%)a, b
Control mediumc	100.00
Salts (0.2% w/v):	
NaN03	145.00
KN03	163.80
MgS04	237.70
CaCl2	198.80
FeS04	228.57
(NH4)2S04	103.60
NH4Cl	72.41
Yeast extract (0.5%)	60.21

[a]Specific activity (U/mg): calculated as µg galacturonic acid $.min^{-1}$. mg $protein^{-1}$.

[b]All relative specific activities of PGase were related to U/mg PGase obtained upon using citrus pectin –based basal medium as production medium.

[c]Citrus pectin based- basal medium composition as mentioned in materials and methods.

Embaby et al.

Embaby et al. SpringerPlus 2014 3:327, doi:10.1186/2193-1801-3-327

Concerning the effect of agitation speed on PGase production, three different agitation speeds (100, 150 and 200 rpm) were studied. The level of PGase obtained upon conducting the fermentation process at 150 rpm was higher than that obtained upon conducting the process at 100 rpm (Data not shown). Whereas, the level of PGase obtained upon conducing the fermentation process at 150 rpm was comparable with that obtained upon conducting the process at 200 rpm (Data not shown). Accordingly, agitation speed at 150 rpm was selected to carry out further optimization experiments.

PBD Results

PBD matrix, coded-real values of independent variables and experimental vs. predicted values of PGase were shown in Table 3. Whilst regression analysis and independent variables evidenced significant consequences on PGase levels were presented in Table 4. The detected PGase activity ranged from 0.0 to 3.04 U/ml reflecting the irreplaceable necessitate for carrying out optimization in order to attain the possible highest levels of PGase. ANOVA results showed that the P-value and F-value of the model were 0.00087 and 10.26, respectively. This F-value 10.26112 of the model reflects the significance of the model. However, the model P-value implies that the chance is only 0.087% that this model F-value could occur due to noise. Values of "Prob. > *F*" less than 0.05 were taken into account to have significant substantial effect on the outcome. Generally, significance of the coefficients has been reported to be directly proportional to t-test and inversely to P-value (Douglas 2001; Heck et al. 2005). Regression analysis suggested that the level of PGase was significantly affected by only three independent variables out of ten tested independent variables. These three independent variables showing significant effects at $P < 0.05$ were orange peel waste percent, pH of the production medium and incubation temperature. Pareto chart (Figure 1) is a convenient way to illustrate the order of significance of independent variables affecting PGase production based on their P-values. After exclusion of the insignificant model terms (based on their insignificant P-values >0.05), a modified first order polynomial equation (1) was set in terms of coded independent variables in order to describe the linear effects of orange peel waste percent, pH of the production medium and incubation temperature on PGase level.

Table 3: PBD with coded levels of ten independent variables and experimental vs. predicted PGase values

Trial#	X1	X2	X3	X4	X5	X6	X7	X8	X9	X10	Y (PGase) (U/mL)c	
											Expa.	Predb.
1	1	-1	1	1	1	1	-1	-1	1	1	0.00	0.1585
2	-1	1	-1	1	1	1	1	-1	-1	1	1.60	1.6135
3	-1	-1	1	-1	1	-1	1	1	1	1	0.64	1.0695
4	1	1	-1	-1	1	1	-1	1	1	-1	0.15	0.0195
5	1	1	1	1	-1	-1	1	1	-1	1	2.70	2.3865
6	-1	-1	-1	1	-1	1	-1	1	1	1	0.00	-0.5795
7	1	-1	1	-1	1	1	1	1	-1	-1	3.04	2.6755
8	-1	1	-1	1	-1	1	1	1	1	-1	0.62	1.0835
9	-1	1	1	-1	-1	-1	-1	1	-1	1	0.00	0.1175
10	-1	-1	-1	-1	1	-1	1	-1	1	1	0.78	0.5935
11	1	1	-1	-1	-1	-1	1	-1	1	-1	1.04	1.0895
12	-1	-1	-1	-1	-1	-1	-1	-1	-1	-1	0.24	-0.0305
13	1	-1	-1	-1	-1	1	-1	1	-1	1	0.00	0.4895
14	-1	1	1	1	1	-1	-1	1	1	-1	0.15	-0.0145
15	1	1	1	-1	-1	1	1	-1	1	1	1.37	1.1425
16	-1	-1	1	1	-1	1	1	-1	-1	-1	1.94	2.0235
17	1	-1	-1	1	1	-1	1	1	-1	-1	2.58	2.6325
18	1	-1	1	1	-1	-1	-1	-1	1	-1	0.00	0.1875
19	-1	1	1	-1	1	1	-1	-1	-1	-1	0.27	0.3635
20	1	1	-1	1	1	-1	-1	-1	-1	1	0.49	0.5885

[a]Experimental values, bpredicted values, cU/mL: calculated as μg galacturonic acid. $min^{-1}.mL^{-1}$.

X_1: orange peel waste, X_2: $NaNO_3$, X_3: $MgSO_4$, X_4: $CaCl_2$, X_5: $FeSO_4$, X_6: KNO_3, X_7: pH, X_8: inoculum size, X_9: incubation temperature and X_{10}: incubation time.

Embaby et al.

Embaby et al. SpringerPlus 2014 3:327, doi:10.1186/2193-1801-3-327

Table 4: Real-coded values of ten independent variables and regression analysis for PGase optimization by PBD

Variable	Low level	High level	Main effect	-coefficient	P-value	t-value	% confidence
	-1	+1					
Orange peel waste (% w/v)	0.5	2.0	0.513	0.2565	0.020839*	2.796274	97.92
NaNO3 (% w/v)	0.1	0.4	-0.083	-0.0415	0.661675	-0.45242	33.84
MgSO4 (% w/v)	0.1	0.4	0.261	0.1305	0.188555	1.422666	81.15
CaCl2 (% w/v)	0.1	0.4	0.255	0.1275	0.197956	1.389961	80.21
FeSO4 (% w/v)	0.1	0.4	0.179	0.0895	0.354709	0.975698	63.53
KNO3 (% w/v)	0.1	0.4	0.037	0.0185	0.844651	0.201681	15.45
pH	6.0	9.0	1.501	0.7505	1.85E-05*	8.181689	99.99
Inoculum size (% v/v)	1.0	5.0	0.215	0.1075	0.271314	1.171928	72.87
Incubation temperature °C	37.0	50.0	-0.811	-0.4055	0.00167*	-4.42062	99.83
Incubation time (hrs)	24.0	60.0	-0.245	-0.1225	0.214517	-1.33545	78.55

*Significant P-value <0.05 & $R^2 = 0.93$ & Adjusted $R^2 = 0.83$ and P -value for the model = 0.000869.

Embaby et al.

Embaby et al. Springer Plus 2014 3:327, doi:10.1186/2193-1801-3-327

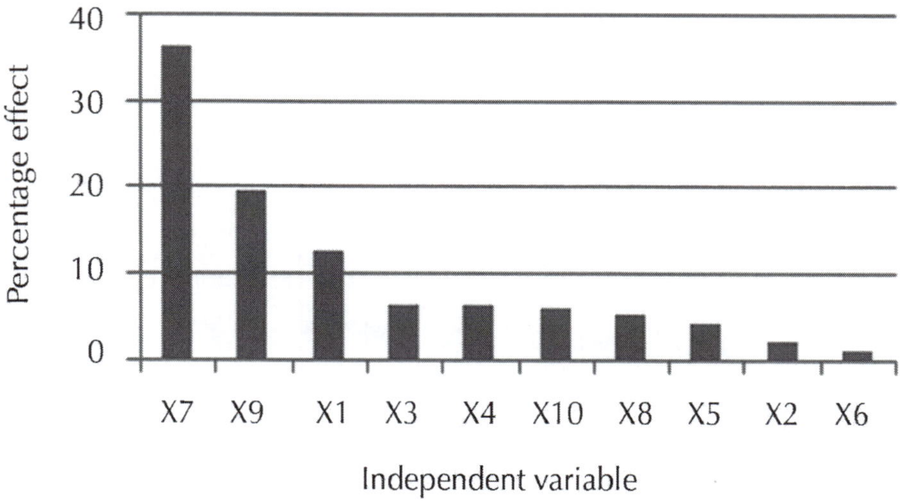

Figure 1: Pareto-Chart in descending layout for PB parameter estimates of ten tested independent variables.

$$Y = 0.881 + 0.257X_1 + 0.751X_7 - 0.406X_9$$

These three chosen independent variables recognized by PB were considered to be the main significant key determinants for PGase production by *B. licheniformis* SHG10. They were further studied in the next stage of the optimization plan via RSM.

RSM Results

Box-Behnken design, a kind of RSM approach, was employed in this study in order to locate the optimal level of the three independent key determinants identified through PB that controlling PGase production. The design matrix along with the experimental and predicted PGase levels were displayed in Table 5. ANOVA results showed that the model F-value of 24.3 and the model P-value of 0.0013 imply the significance of the model and the likelihood (0.13%) that this F-value could occur due to noise. Moreover, adequacy of the model to explain the relationship between the response (output) and the significant independent variables could be measured by the small model P-value 0.0013 and the large lack of fit P-value of 0.276. The lack of fit F-value (2.77) indicates that it is not significant relative to the pure error. Non-significant lack of fit reflects goodness of the model. Aptness of the model was inferred from the R^2 value of 0.977. Regression coefficients were calculated in terms of coded values of independent variables and data were fitted to a second order polynomial equation (2).

Table 5: Box-Behnken design of three independent variables and experimental vs. predicted PGase values

Trial#	X1	X7	X9	Y (PGase) (U/mg)c	
				Expa.	Predb.
1	-1	-1	0	0.85	0.767
2	1	-1	0	2.03	2.052
3	-1	1	0	2.27	2.247
4	1	1	0	1.15	1.232
5	-1	0	-1	1.79	1.708
6	1	0	-1	2.22	2.033
7	-1	0	1	0.88	1.066
8	1	0	1	0.93	1.011
9	0	-1	-1	0.823	0.987
10	0	1	-1	1.17	1.273
11	0	-1	1	0.215	0.111
12	0	1	1	0.65	0.485
13	0	0	0	2.48	2.640
14	0	0	0	2.68	2.640
15	0	0	0	2.76	2.640

[a]Experimental values.

[b]Predicted values.

[c]U/mg: calculated as μg galacturonic acid.min^{-1}.mg^{-1}.

X_1: orange peel waste percent, X_7: pH of the production medium and X_9: incubation temperature.

Embaby *et al.*

Embaby *et al. SpringerPlus* 2014 **3**:322, doi:10.1186/2193-1801-3-327

$$Y = 2.64 + 0.0675X_1 + 0.165X_7 - 0.416X_9 - 0.1622X_1^2 - 0.9028X_7^2 - 1.022X_9^2 - 0.575X_1X_7 - 0.095X_1X_9 + 0.022X_7X_9 \quad (2)$$

Our data revealed that only four out of nine model terms exhibited significant effect ($P < 0.05$) on PGase production (Table 6). The independent variable incubation temperature showed both linear and quadratic effect at P-values of 0.0023 and 0.00022, respectively. Dissimilarly, the independent variable pH of production medium exhibited both linear and cross interacted effect with the independent variable orange peel waste at P- values of 0.00039 and 0.0026, respectively.

Table 6: Real–coded values of independent variables and regression analysis for PGase optimization by Box-Behnken

Variable	Low level	Middle level	High level	Model term	Main effect	-coefficient	t-value	P-value	% confidence
	-1	0	+1						
Orange peel waste (w/v %)	0.5	1.5	2.5	X1	0.135	0.0675	0.92151	0.399067	60.10
pH	6.0	8.0	10.0	X7	0.3306	0.16525	2.255994	0.073721	92.63
Incubation temp. °C	30.0	40.0	50.0	X9	-0.832	-0.416	-5.67923	0.002357*	99.76
				X12	-0.325	-0.16225	-1.50482	0.192706	80.73
				X72	-1.805	-0.90275	-8.37274	0.000398*	99.96
				X93	-2.044	-1.02275	-9.48571	0.00022*	99.99
				X1.X7	-1.15	-0.575	-5.55072	0.002608*	99.73
				X1.X9	-0.19	-0.095	-0.91708	0.401174	60.00
				X7.X9	0.044	0.022	0.212375	0.840201	16.00

*Significant P-value <0.05 & $R^2 = 0.977$ & Adjusted $R^2 = 0.93$ and P -value for the model = 0.0013.

Embaby *et al.*

Embaby *et al.* *SpringerPlus* 2014 3:327, doi:10.1186/2193-1801-3-327

In order to attain the optimized conditions, canonical analysis was carried out. Studying the overall shape of the response and determining whether the stationary point is maximum, minimum or saddle point could be achieved through canonical analysis. Shape of the response is characterized by eigen-values and eigenvectors in the matrix of second order. Directions of principle orientation for the surface are determined by eigenvectors, while signs and magnitude of eigen-values point for surface shape in these directions. Two rules of thumb explaining the concept of eigen-values and their mathematical indications were reported previously (Myers 1976). The 1st rule states that upward and downward curvatures of the response are evidenced by positive and negative eigen-values, respectively. While the 2nd rule states that the larger an eigen–value is in its absolute value, the more pronounced is the curvature of the response surface in the associated direction. Our data revealed that, the model has eigen-values of [$_1$=-0.06128758, $_7$=-1.00026881 and $_9$=-1.02619361]. By applying 1st rule of Myer, our negative eigen- values reflected that the predicted stationary point is maximum. However, based on 2nd rule of Meyer, the two largest eigen-values in their absolute values (1.00026881 and 1.02619361) of our model conferred a pronounced curvature in the directions of two independent variables (X_7 and X_9). This finding to a great extent authenticated the results of regression analysis stated that X_7 and X_9 exhibited the highest significant effect in linear, quadratic and cross interacted forms on PGase level. Anchored in canonical analysis, the predicted coded stationary point was at {X_1=0.260620063, X_7=0.005901057 and X_9=-0.215413876} to achieve a predicted Y of 2.69 μg galacturonic acid.min^{-1}. mg^{-1}. Besides, the predicted stationary point is clearly positioned inside the explored domain (model constrains).

To further explore the nature of the response surface at the stationary point, three dimensional contour surface plots were generated. Figures 2, 3 and 4 illustrated the contour surface plots for the response. Typically, the contour surface plots are based on the model, holding one independent variable constant at its optimal level where varying the other two independent variables within the domain. Figure 2 illustrated the response of the dependent variable (PGase) for the optimal level of incubation temperature. The maximal predicted level of PGase 2.69 μg galacturonic acid.min^{-1}. mg^{-1} was noticed at levels of 1.76% (w/v) and 8.0 for orange peel waste and pH of the production medium, respectively nearby to the center point of the model. However, the contour surface plot depicted in Figure 3 revealed that the maximal point of PGase at the optimal level for pH of the production medium could be reached at 1.76% (w/v) orange peel waste and 37.8°C. Correspondingly, these predicted levels of both dependent and the independent variables were further evidenced by the contour surface plot illustrated in Figure 4. Concentrations of orange peel waste greater than 1.7% (w/v) did not result in further enhancement in the level of PGase as it was revealed form contour surface plots depicted in Figures 2 and 3. This reflects that a stationary point is achieved at concentrations of orange peel waste exceeding 1.76% (w/v). Conversely higher levels of pH of the production medium and incubation temperature beyond 8.0 and 37.8°C, respectively showed adverse effect on the level of PGase as illustrated from the contour surface plot depicted in Figure 4. These results were verified by the above results of canonical analysis regarding sign and magnitude of model eigen–values.

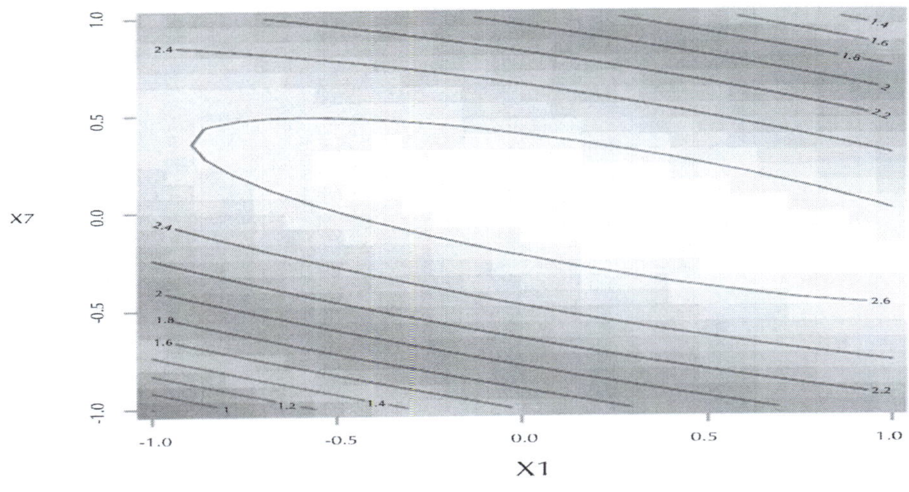

Figure 2: Contour surface plot for the dependent variable PGase vs. the independent variables; orange peel waste and pH of the production medium.

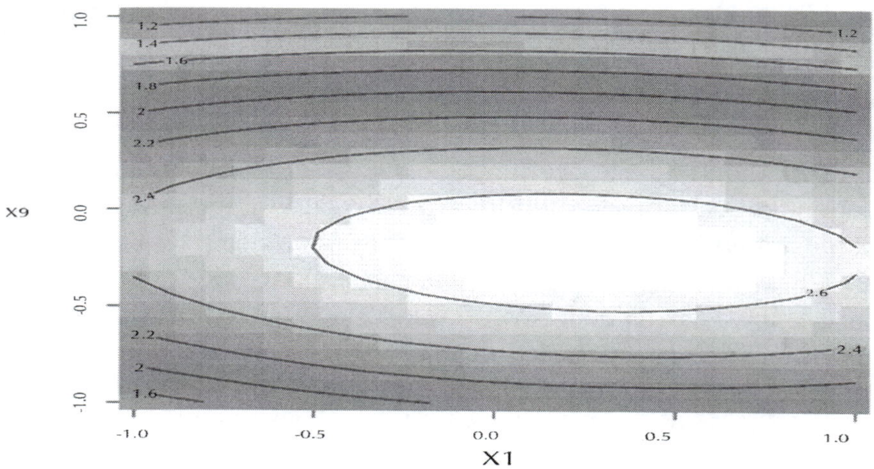

Figure 3: Contour surface plot for the dependent variable PGase vs. the independent variables; orange peel waste and incubation temperature.

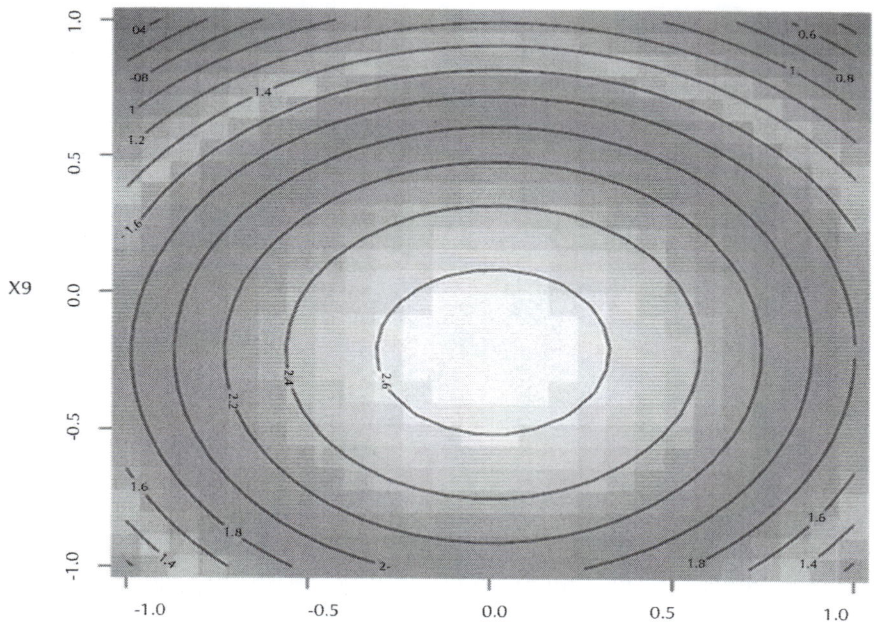

Figure 4: Contour surface plot for the dependent variable PGase vs. the independent variables; pH of the production medium and incubation temperature.

In addition, validation of the model for PGase production was carried out experimentally by using the aforementioned predicted levels of the independent variables. Experimental data revealed that the adequacy of the model was more or less 100%.

Optimum pH and Temperature for Crude PGase

Data revealed that PGase SHG10 showed appreciable level of activity over a wide range of pH (3.0-11.0) (Figure 5a). Whilst, the optimum temperature of enzyme activity was found to be at 50°C (Figure 5b).

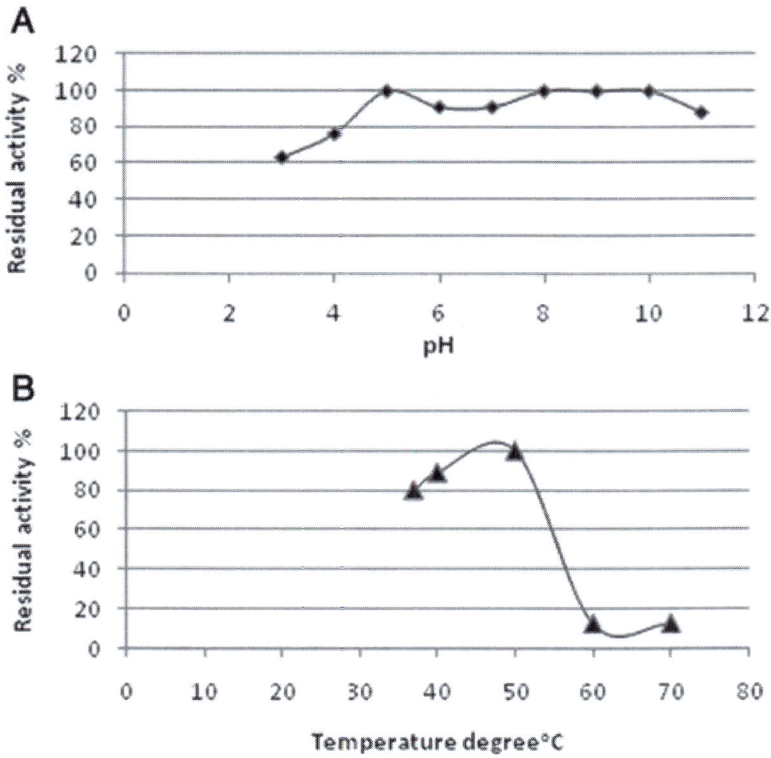

Figure 5: Effect of different pH(s) [A] and temperature degrees [B] on the activity of PGase SHG10.

Levels of Pectic Oligosaccharides

The level of pecic oligosaccharides accumulated in the fermentation broth of this bioprocess was estimated. It was found to be 200 µg galacturonic acid/mL of fermentation broth after 24 hrs. No higher levels of these substances were found in the fermentation broth beyond 24 hrs of incubation.

DISCUSSION

PGase(s), one member of the family pectinase(s) that have a vast number of industrial applications, still attract the attention of many researchers worldwide. Due to the potential and wide applications of pectinases particularly PGase(s), researchers up till now in worldwide laboratories report the isolation and characterization of novel PGase(s) mainly from fungi and barely from bacteria. Current commercial PGase preparations available in enzymes markets are exclusively derived from fungal species. Normally, PGase (s) of fungal origin have an optima pH ranging from 3.5-5.5. This restricted acidic range of optimal pH greatly confines the extent of fungal PGase (s) utilization. Commercialization of any new enzyme industry is usually constricted to the high cost of production process. Nonetheless, the cost encounter in culture medium accounts for 30-40% of the overall production cost (Bayoumi et al. 2008). From another side, the non-efficiently utilized agro-industrial wastes are accumulated in considerable amounts that could not be abandoned. Furthermore, applying of improper waste disposal methods results in arise of health hazards and environmental problems. Here, the process of PGase production directed by *B. licheniformis* SHG10 was studied thoroughly from the standpoint of low cost effectiveness in conjunction with appreciable yield.

In the course of search for low cost effective medium composition to simultaneously support the growth of *B. licheniformis* SHG10 and induce PGase production, different raw agro-industrial wastes (orange, lemon, pomegranate, banana and artichokes peel wastes) had been introduced separately into the fermentation broth as sole PGase inducers. Our data demonstrated that there exists a profound impact imposed by carbon source on the production of PGase from *B. licheniformis* SHG10. Present finding reveals that raw orange peel waste is a superior PGase inducer comparable with other raw agro-industrial wastes, synthetic citrus pectin and other synthetic carbon sources being tested in this study. Present finding is in disagreement with that of Dey et al. (2011) who reported that synthetic citrus pectin exhibited the best effect as PGase inducer from *Bacillus* sp. AD1 relative to some agro-industrial wastes particularly lemon and orange peel wastes. It was reported that synthetic citrus pectin is the best PGase inducer from *E. aerogenes* NBO2 (Darah et al. 2013). In accordance with our findings, orange peel waste was the best inducer of PGase from *Aspergillus niveus* in submerged state fermentation (Maller et al.2011). With regard to the impact of wheat bran on PGase productivity from our bacterial strain, data revealed that almost quite similar levels of PGase were obtained upon adding synthetic citrus pectin and wheat bran separately in the fermentation broth. In this respect, the finding of Rehman et al. (2012) is in accordance with our findings. Regarding the impact of synthetic carbon sources alternative to citrus pectin as PGase inducers, none of the tested synthetic carbon sources exhibited stimulatory effect on PGase productivity by *B. licheniformis* SHG10. Findings of Rehman et al. (2012); Jayani et al. (2010) in relation to introduction of glucose, maltose, sucrose, fructose and glycerol into the fermentation broth of *B. licheniformis* KIBGE IB-21 and *B. sphaericus*(MTCC7542), respectively are in good agreement with our finding. Moreover, glucose, xylose, maltose and sucrose as sole PGase inducers were reported to exhibit inhibitory effect on PGase production from *B.firmus* -I-10104 in solid state fermentation (SSF). Glucose and sucrose at a concentration of 1% (w/v) was pointed out to inhibit the PGase productivity from *E. aerogenes*NBO2 (Darah et al. 2013). The noticeable reduction in PGase levels in presence of sugars as a sole carbon source and a PGase inducer could be attributed to the phenomenon of catabolite repression (Ahlawat et al. 2009; Cavalitto et al. 1996; Solís-Pereira et al. 1993).

It was reported that productivity of bacterial and fungal PGase(s) is greatly affected by some mineral salts added to the fermentation broth. To explore the role of these mineral salts on PGase productivity by *B. licheniformis* SHG10, the effect of some mineral salts mentioned above was investigated. Present data revealed that, introduction of $MgSO_4$, $FeSO_4$, $CaCl_2$, KNO_3 and $NaNO_3$at a concentration of 0.2% (w/v) to the fermentation broth containing synthetic citrus pectin as co-inducer sole carbon source resulted in variable higher appreciable levels of

PGase comparable to those levels obtained at zero concentrations of these salts. In accordance with the present finding, KNO_3 at a concentration of 0.2% (w/v) was the best PGase nitrogen source inducer form *B. licheniformis* growing on the agro-industrial potato peels waste in SSF(Dharmik and Gomashe2013). In contrast to our finding, neither $NaNO_3$ nor KNO_3 at a concentration of 0.1% (w/v) had shown stimulatory effects on PGase productivity from *B. sphareicus* (MTCC 7542) in citrus pectin containing fermentation broth (Jayani, et al. 2010). Similarly, $NaNO_3$ exhibited an inhibitory effect on PGase production from *E. aerogenes* NBO2 (Darah et al. 2013). However, other mineral salts such as NH_4Cl and $(NH_4)_2SO_4$ exhibited inhibitory effect on PGase from *B.licheniformis* SHG10. Addition of these two mineral salts separately to the fermentation broth displayed variable levels of PGase from different PGase producers either higher or lower than those obtained at zero concentrations of each salt (Jayani, et al. 2010; Darah et al. 2013; Bayoumi et al. 2008; Dharmik and Gomashe 2013; Kashyap et al. 2003). Concerning yeast extract, also the literature reported varied impact on PGase production from different PGase producers (Rehman et al. 2012; Kapoor et al. 2000).

Low constitutional levels of PGase from *B. licheniformis* SHG10 in presence of peptone, beef or tryptone were detected in fermentation broth lacking any source of pectin either in synthetic form or in raw one. However, other studies reported varied levels of PGase from different PGase producers in fermentation broth containing peptone or tryptone as additional nitrogen sources in presence of pectin source.

In the course of high yield PGase production from *B. licheniformis* SHG10, statistical optimization was applied to maximize the yield of PGase. Roughly identified factors through OVAT, showing a stimulatory effect on PGase production from *B.licheniformis* SHG10 growing on citrus pectin based-basal medium, were studied further through PB. Statistical analysis of data derived from PB revealed that none of the tested mineral salts identified through OVAT; $MgSO_4$, KNO_3, $NaNO_3$, $CaCl_2$ and $FeSO_4$; showed significant impact on PGase productivity in fermentation broth containing orange peel waste instead of synthetic citrus pectin. The agro-industrial orange peel waste seems to be a rich substrate that could co-simultaneously provide the bacterium with all elements needed during the course of bacterial growth and PGase induction as well. This finding greatly alleviates the cost of the PGase production medium.

Substitution of high cost synthetic substrates with raw solid substrates such as agro-industrial wastes in enzyme industry based-bioprocesses is somehow feasible task implying some challenges. The prospect or the scope that a solid waste could cover all needs of a microorganism from organic and inorganic substances mandatory for microbial growth and enzyme induction is considered to be one of the major obstacles in this respect. The greater that essential nutritional elements, mandatory for microbial growth and enzyme induction, exist in suboptimal levels in raw solid substrates, the more indispensable need is to add external supplements in the fermentation broth (Sneath 1986). As a consequence, the cost of the production medium will elevate.

Optimized production of PGase from *B. licheniformis* SHG10 is achievable through using a low concentration of orange peel waste (1.76% w/v) in the fermentation broth as the sole carbon source and the only PGase inducer. This could be attributed to the high content of pectin included in this waste. From another side, the need for low percentage of this waste to support PGase production from *B. licheniformis* SHG10 implies the feasibility of the downstream process in order to remove the remaining undegraded orange peel waste. As a consequence the cost encountered in downstream process would decrease and the overall cost of the process of PGase production would also reduce.

Pertaining to optimal temperature and pH of the production medium required for PGase production, our finding was in a good agreement and disagreement with other findings (Soares et al. 1999; Dey et al. 2011; Maller et al. 2011; de Andrade et al. 2011; Das et al. 2011; Deshmukh et al.2012). In the context of optima pH for PGase activity, the crude PGase SHG10 was observed to work efficiently under a wide range of pH covering from 3.0 to 11.0 with a slight noticeable decline in the activity at pH 11. Whilst, the optimal activity was confined to the neutral-alkaline scale. In addition, the crude PGase SHG10 showed its activity over a wide range of temperatures ranging from 37°C – 50°C with an optimal activity at 50°C. The requirement of alkaline pH range (8.0 - 11.0) for fulfillment of PGase SHG10 optimal activity features the potential applications of this enzyme in textile processing, degumming of plant bast fibers, treatment of pectic wastewaters, paper making, and coffee and tea fermentations. These two findings concerning optima pH and temperature of our PGase SHG10 imply a clue about the wide range of pH and temperature degrees under which SHG10 PGase could work efficiently. Whereupon, PGase SHG10 would have vast industrial promising applications where each selected combination of pH and temperature degree is a crucial factor to guarantee the success of a certain industrial application. As a consequence, PGase SHG 10 would have a privilege over fungal PGase (s) that have a restricted range of both optima pH and temperature degrees.

Regarding the issue of pectic oligosaccharides production, the literature contains a plethora of chemical methods devoted to synthesize the pectic oligosaccharides that have some reported medical applications (Olano-Martin et al. 2003a,[b]). In this study, data revealed the likelihood of obtaining an appreciated high level of these substances (200 μg galacturonic acid/mL) in the fermentation broth of *B. licheniformis* SHG10 growing on the agro-industrial orange peel waste as a sole carbon source. These recorded levels of pectic oligosaccharides resulted as a consequence of the activity of the PGase produced in the fermentation broth of *B. licheniformis* SHG10 on its complex substrate, orange peel waste. As a matter of fact, the more PGase produced is in the fermentation broth the more liberated pectic oligosaccharides would exist. Therefore, this greatly would necessitate the need for optimizing the process of orange peel waste biodegradation in terms of maximizing pectic oligosaccharides production in the future. However, the nature of the obtained pectic oligosaccharides could be controlled through the possible cascade of pectinase(s) that would be produced by *B. licheniformis* SHG10 growing under the stated conditions. This finding is considered an alternative promising approach implying some challenges towards biosynthesis of pectic oligosaccharides from the standpoints of cost effectiveness, good quality and high yield.

CONCLUSIONS

Present work addresses a cheap rapid biotechnological method to promote PGase and pectic oligosaccharides industries through bioprocessing of the agro-industrial orange peel waste via *B. licheniformis* SHG10. Characteristics of PGase SHG10 concerning the wide range of optima pH greatly confirm its potential biotechnological applications. For the next future work, the authors are planning to further maximize the yield of PGase from *B. licheniformis* SHG10 through a molecular gene cloning approach and improve the physicochemical properties and PGase activity of the enzyme towards pectin-containing substrates through applying directed evolution methodologies. Additionally, the authors are going to optimize the yield of pectic oligosaccharides accumulated in the fermentation broth as a result of the biodegradation of orange peel waste through this bioprocess.

METHODS

Bacterial Strain

Bacillus licheniformis strain SHG10 was used in this study as PGase producer. This bacterium was previously isolated from Egyptian soil and identified as *B. licheniformis* SHG10 strain (unpublished data). Its 16S rDNA nucleotide sequence was submitted in the GenBank at NCBI [National Center for Biotechnology Information] under the accession number [GenBank: JN853580]. Moreover, this bacterial strain was deposited in the DSMZ [Leibniz Institute-German Collection of Microorganisms and Cell Cultures] under the accession number [DSM 28096].

Pectin-Containing Materials

Different pectin-containing materials were co-utilized in this study as a sole carbon source for the growth of the producer bacterial strain and as an inducer for PGase as well. Pectin-containing materials used in this study included synthetic citrus pectin, wheat bran, orange peel waste, lemon peel waste, banana peel waste, artichokes peel waste and pomegranate peel waste. The last five pectin–containing materials were collected from different sites (e.g.; local Egyptian markets, domestic effluents and agricultural fields). The collected pectin–containing materials were washed with distilled water then were allowed to dry at 60°C for five hrs. After that, the dried pectin–containing materials were prepared in the form of small cut pieces preparation before their incorporation into the fermentation broth. Synthetic citrus pectin was purchased from Sigma –Aldrich Co. While, wheat bran was obtained from flour mills companies in Alexandria, Egypt.

Media

Peptone yeast broth (Bernhardt et al. 1978) was used to activate the bacterial producer strain. PA medium is peptone yeast broth with 1.5% agar agar. Polygalacturonase core production medium of (Soares et al. 1999) with slight modifications was used in this study. This modified medium was used during the initial steps of the optimization process. The modified PGase core production medium contained the following components in % (w/v): 0.5 g pectin, 0.14 g $(NH_4)_2SO_4$, 0.6 g K_2HPO_4, 0.2 g KH_2PO_4, 0.01 g $MgSO_4$ and 0.3 g yeast extract unless otherwise stated.

Inoculum Preparation

A fine touch of *B. licheniformis* SHG10 preserved on PA slant was streaked on PA agar and was incubated at 37°C for overnight. One colony was picked to inoculate 20 ml of PY medium in 100 ml Erlenmeyer flask. The inoculated broth was incubated at 37°C with agitation speed of 200 rpm for 4 hrs until the culture OD at 420 nm reached 0.5. Then this growing culture (seed broth) was used to inoculate the fermentation broth (production medium). The inoculum size of the seed broth used to inoculate the fermentation broth was 2%(v/v) unless otherwise stated.

PGase Assay

The PGase activity was assayed by estimating the amount of reducing sugars released under assay conditions. Determination of the amount of released reducing sugars as galacturonic acid was carried out as reported previously (Miller 1959) using 2-hydroxy-3,5-dinitrobenzoic acid [(DNSA), Shanghai Orgpharma Chemical Co., Ltd., China]. Succinctly, the reaction mixture contained 0.5 mL of 0.5% citrus pectin (Sigma-Aldrich Co.) as a substrate (dissolved in 50 mM Tris-HCl, pH7.6) and 0.5 mL crude enzyme (fermentation broth). This mixture was incubated at 37°C for 20 min. After that, the enzymatic reaction was stopped by addition of 1 mL of DNSA followed by boiling for 10 min. Then the final volume was completed to 4 mL by distilled water and the developed color was measured at 540 nm. Control reactions were prepared as mentioned above except that DNSA was added prior addition of the crude enzyme. A standard curve with -galacturonic acid (Sigma-Aldrich Co.) was established. One unit (arbitrary unit) of enzyme activity was defined as the amount of enzyme that releases one μg of -galacturonic acid per min from citrus pectin as a substrate in 50 mM Tris-HCl, pH 7.6 at 37°C.

Protein Determination

The protein content of the crude enzyme solution was performed as reported previously using Folin -Lowry reagent (Lowry et al. 1951). A standard curve using bovine serum albumin was established.

Pectic Oligosaccharides Determination

Pectic oligosaccharides were determined as reported previously by the method of Miller (1959). Briefly, 0.5 mL of the fermentation broth was added to 1 mL of DNSA. Then, the reaction mixture was boiled for 10 min. After that, the absorbance of the developed color was measured at 540 nm against blank (the same as the reaction except that water was added instead of fermentation broth).

Optimizing the Production of PGase from *B. Licheniformis* SHG10

Optimizing the production of PGase from the producer strain was accomplished through a three successive step plan; one variable at a time approach (OVAT), Plackett-Burman design and Box- Behnken design.

OVAT Approach

OVAT was employed in this study in order to screen different independent variables that would either stimulate or inhibit the production of PGase enzyme. This approach is based on changing one variable at a time without studying the interaction among the tested variables. The effect of different agro-industrial wastes (e.g., orange, lemon, banana, artichoke, pomegranate peel wastes and wheat bran) and some synthetic carbon sources (e.g., citrus pectin, tryptone, peptone, beef extract, glucose, maltose, sucrose, xylose, fructose and glycerol) on PGase productivity by *B. licheniformis* was assessed. In addition, the effect of different salts such as $NaNO_3$, KNO_3, NH_4Cl, $CaCl_2$, $MgSO_4$ and $FeSO_4$ was studied as well. The exact concentrations of the aforementioned tested substances were displayed in Tables 1 and 2.

Plackett – Burman Design (PBD)

Identifying the significant main key determinants (physicochemical independent parameters) in a bioprocess along with studying the linear effect of these tested variables is achieved by applying a powerful statistical approach namely called Plackett–Burman design that developed by two statisticians Plackett and Burman (1946). In this approach evaluating the linear effect of *N* independent variables on the dependent variable (output of a bioprocess) is tested in a *N + 1* experiment. Normally, each independent variable is studied in two levels -1 and +1; low coded level and high coded level, respectively. The design matrix was generated by a statistical software package Minitab version 15. Here, twenty experimental runs (trials) had been conducted. The following polynomial equation from the first order (Equation 3) was put in order to evaluate the linear effect imposed by the ten tested independent variables on the level of PGase enzyme:

$$Y = \beta_0 + \beta_1 X_1 + \beta_2 X_2 + \beta_3 X_3 + \beta_4 X_4 + \beta_5 X_5 + \beta_6 X_6 + \beta_7 X_7 + \beta_8 X_8 + \beta_9 X_9 + \beta_{10} X_{10} \quad (3)$$

Where *Y* is the level of PGase activity, $_0$ is the model intercept, X_1-X_{10} are the tested independent variables (orange peel waste, $NaNO_3$, $MgSO_4$, $CaCl_2$, $FeSO_4$, KNO_3, pH, inoculum size, incubation temperature and incubation time, respectively) and $_1$- $_{10}$ are the coefficient of the ten tested independent variables. The experimental runs were conducted according to the PBD matrix in 250 ml Erlenmeyer flasks with working volume of 25 ml. All experimental runs were conducted at agitation speed of 150 rpm.

Box-Behnken Design

Three key determinants (orange peel waste percent, pH of the production medium and incubation temperature) identified through PBD had significant effects on the level of PGase. In order to determine the optimal level of each key determinant ((independent variable) along with the maximal level of PGase (dependent variable), a response surface methodology approach was applied here. Box- Behnken design, developed by Box and Behnken (1960), was employed in this study. Fifteen experimental runs (trials) had been conducted. The following polynomial equation from the second order (Equation 4) was set in order to estimate the effect of all possible forms of interactions imposed by the above mentioned three independent variables on the level of PGase enzyme:

$$Y = \beta_0 + \beta_1 X_1 + \beta_7 X_7 + \beta_9 X_9 + \beta_{11} X_1^2 + \beta_{77} X_7^2 + \beta_{99} X_9^2 + \beta_{17} X_1 X_7 + \beta_{19} X_1 X_9 + \beta_{79} X_7 X_9 \quad (4)$$

Where *Y* is the level of PGase activity, $_{0}$ is the model intercept, X_1, X_7 and X_9 are the tested independent variables (orange peel waste percent, pH of the production medium, and incubation temperature, respectively), $_{1}$, $_{7}$ and $_{9}$ are linear coefficients, ($_{11}$, $_{77}$, $_{99}$) are quadratic coefficients and ($_{17}$, $_{19}$, $_{79}$) are cross interaction coefficients. For statistical calculations, each independent variable X was coded as Xi according to the Equation 5.

$$X_i = (x_i - x_o) / \Delta xi \quad (5)$$

Where X_i is dimensional coded value of the independent variable, x_i is the real value of this variable at this coded value; x_o is the real value of this variable at the center point (zero level) and *xi* is the step change value. The experimental runs were conducted according to the Box –Behnken matrix in 250 ml Erlenmeyer flasks with a working volume of 25 ml. All experimental runs were conducted at agitation speed of 150 rpm.

Statistical, Canonical Analyses and Contour Plots

RSM package (R Development Core Team 2009), available from the Comprehensive R Archive Network at http://CRAN.R-project.org/package=rsm, was used in this study to carry out multiple regression, canonical analyses and graphing of three dimensional contour surface plots.

Effect of Different pH and Temperature on the Activity of PGase

Two buffers were used in this study to cover a wide range of pH, 50 mM sodium acetate buffer, pH 3-6 and 50 mM phosphate buffer, pH 7-11. Five different temperatures (37°C, 40°C, 50°C, 60°C and 70°C) were used to test the optimal activity of PGase.

AUTHORS' CONTRIBUTION

AME: She participated in putting the design of most experiments especially those concerned with statistical modeling. She wrote and revised the whole manuscript. She was responsible for statistical analysis, scientific interpretation and discussion of the obtained data. AAM: She was responsible for carrying out the laboratory experiments. HMS: She participated in statistical analysis for the data of Response surface methodology. NZS: She revised the English language of the manuscript. TMG: She participated in analysis and interpretation of the obtained data. All authors read and approved the final manuscript.

REFERENCES

1. Ahlawat S, Dhiman SS, Battan B, Mandhan RP, Sharma J (2009) Pectinase production by *Bacillus subtilis* and its potential application in biopreparation of cotton and micropoly fabric. Process Biochem 44(5):521-526
2. Alkorta I, Garbisu C, Llama MJ, Serra JL (1998) Industrial

applications of pectic enzymes—a review. Process Biochem 33:21-28

3. Balu AM, Budarin V, Shuttleworth PS, Pfaltzgraff LA, Waldron K, Luque R, Clark JH (2012) Valorization of orange peel residues: waste to biochemicals and nanoporous materials. Chem Sus Chem 5(9):1694-1697
4. Basu S, Saha MN, Chattopadhyay D, Chakrabarti K (2009) Large scale degumming of ramie fiber using a newly isolated *Bacillus pumilus* DKS1 with high pectate lyase activity. J Ind Microbiol Biotechnol 36:239-245
5. Bayoumi RA, Yassin HM, Swelim MA, Abdel-All EZ (2008) Production of Bacterial Pectinase(s) from Agro-Industrial Wastes Under Solid State Fermentation Conditions. J Appl Sci Res 4(12):1708-1721
6. Bernhardt K, Schrempf H, Goebel W (1978) Bacteriocin and antibiotic resistance plasmids in*Bacillus cereus* and *Bacillus subtilis*. J Bacteriol 133:897-900
7. Box GEP, Behnken DW (1960) Some new three level design for study of quantitative variables. Technometrics 2:455-475
8. Cavalitto SF, Areas JA, Hours RA (1996) Pectinase production profile of *Aspergillus foetidus* in solid state cultures at different acidities. Biotechnol Lett 18:251-256
9. DarahI, NishaM, LimSH(2013)EnhancementofPolygalacturonase Production from *Enterobacter Aerogenes* NBO2 by Submerged Fermentation. Adv Stud 5(5):173-189
10. Das B, Chakraborty A, Ghosh S, Chkarabartik (2011) Studies on the effect of pH and carbon sources on enzyme activities of some pectinolytic bacteria isolated from jute retting water. Turk J Biol 35:671-678
11. de Andrade MVV, Delatorre AB, Ladeira SA, Martins MLL (2011) Production and partial characterization of alkaline polygalacturonase secreted by thermophilic *Bacillus* sp. SMIA-2 under submerged culture using pectin and corn steep liquor. Ciência e Tecnologia de Alimentos 31(1):204-208
12. Demir H, Gogus N, Tari C, Heerd D, Lahore MF (2012) Optimization process parameters for the utilization of orange peel to produce polygalacturonase by solid-state fermentation from an*Aspergillus sojae* mutant strain. Turk J Biol 36:394-404

13. Deshmukh N, Talkal R, Jha K, Singh PG, Prajapati DC (2012) Production, Purification, Characterization and Comparison of Polygalacturonase from various strains of *Aspergillus*. Int J Sci Technol Res 1(9):85-91
14. Development Core Team (2009) R: a language and environment for statistical computing. R Foundation for Statistical Computing, Vienna, Austria. http://www.R-project.org/
15. Dey A, Karmakar M, Ray RR (2011) Extracellular Thermostable Polygalacturonase from *Bacillus*sp. AD 1. Der Pharm Lett 3(2):358-367
16. Dharmik PG, Gomashe AV (2013) Bacterial polygalacturonase (PG) production from agro industrial waste by solid state fermentation. Ind J Appl Res 3(6):439-442
17. Douglas CM (2001) Design and analysis experiments. Wiley, Arizona, USA.
18. Gummadi S, Panda T (2003) Purification and biochemical properties of microbial pectinases – a review. Process Biochem 38:987-996
19. Heck JX, De Barros S, Ayub M (2005) Optimization of xylanase and mannanase production by*Bacillus circulans* strain BL53 on solid state cultivation. Enzyme Microb Technol 37:417-423
20. Hoondal GS, Tiwari RP, Tewari R, Dahiya N, Beg QK (2002) Microbial alkaline pectinases and their industrial applications: a review. Appl Microbiol Biotechnol 59:409-418
21. Jayani RS, Saxena S, Gupta R (2005) Microbial pectinolytic enzymes: a review. Process Biochem 40(9):2931-2944
22. Jayani RS, Shukla SK, Gupta R (2010) Screening of Bacterial Strains for Polygalacturonase Activity: Its Production by *Bacillus sphaericus* (MTCC 7542). Enzym Res 88(3):251-254
23. Kapoor M, Beg QK, Bhushan B, Dadhich KS, Hoondal GS (2000) Production and partial purification and characterization of a thermo-alkali stable polygalacturonase from *Bacillus* sp. MG-cp-2. Process Biochem 36(5):467-473
24. Kashyap DR, Soni SK, Tewari R (2003) Enhanced production of pectinase by *Bacillus* sp.DT7 using solid state fermentation. Biores Technol 88:51-254
25. Kuhad RC, Kapoor M, Rustagi R (2004) Enhanced production of

an alkaline pectinase from*Streptomyces* sp. RCK-SC by whole-cell immobilization and solid state cultivation. World J Microbiol Biotechnol 20:257-260

26. Lowry OH, Rosebrough NJ, Farr AL, Randall RJ (1951) Protein Measurement with the Folin Phenol Reagent. J Biol Chem 193:265-275
27. Maller A, Damásio ARL, da Silva TM, Jorge JA, Terenzi HF, Polizeli MLT (2011) Biotechnological potential of agro-industrial wastes as a carbon source to thermostable polygalacturonase production in *Aspergillus niveus*. Enzym Res 2011:1-6
28. Martin MA, Siles JA, Chica AF, Martin A (2010) Biomethanization of orange peel waste. Biores Technol 101:8993-8999
29. Martín MA, Fernández R, Serrano A, Siles JA (2013) Semi-continuous anaerobic co-digestion of orange peel waste and residual glycerol derived from biodiesel manufacturing. Waste Manage 33(7):1633-1639
30. Miller GL (1959) Use of dinitrosalicylic acid reagent for determination of reducing sugars. Anal Chem 31:426-428
31. Mohamed SA, Drees EA, El-Badry MO, Fahmy AS (2010) Biochemical Properties of α-Amylase from Peel of *Citrus sinensis* cv. Abosora. Appl Biochem Biotechnol 160:2054-2065
32. Möller M, Schmitz PM, Thiele H, Wronka T (2001) Economically and ecologically integrated valuation of land-use in less favored areas|[Integrierte konomische und kologische Bewertung der Landnutzung in peripheren Regionen]. Ber Landwirtsch 79(1):19-48
33. Myers RH (1976) Response Surface Methodology. Edwards Brothers, Ann Arbor, MI.
34. Olano-Martin E, Rimbach GH, Gibson GR, Rastall RA (2003) Pectin and pectic-oligosaccharides induce apoptosis in *in vitro* human colonic adenocarcinoma cells. Anticancer Res 23:341-346
35. Olano-Martin E, Williams MR, Gibson GR, Rastall RA (2003) Pectins and pectic-oligosaccharides inhibit *Escherichia coli* O157:H7 Shiga toxin as directed towards the human colonic cell line HT29. FEMS Microbiol Lett 218:101-105

36. Pedrolli DB, Monteiro AC, Gomes E, Carmona EC (2009) Pectin and pectinases: production, characterization and industrial application of microbial pectinolytic enzymes. Open Biotechnol J 3:9-18

37. Plackett RL, Burman JP (1946) The design of optimum multifactorial experiments. Biometrica 37:305-325

38. Quattara HG, Koffi BL, Karou GT, Sangare A, Niamke SL, Diopoh JK (2008) Implication of *Bacillus*sp. in the production of pectinolytic enzymes during cocoa fermentation. World J Microbial Biotechnol 24:1753-1760

39. Rehman HU, Qader SAU, Aman A (2012) Polygalacturonase: Production of pectin depolymerising enzyme from *Bacillus licheniformis* KIBGE IB-21. Carb Poly 90:387-391

40. Rivas B, Torrado A, Torre P, Converti A, Domínguez JM (2008) Submerged citric acid fermentation on orange peel autohydrolysate. J Agric Food Chem 56(7):2380-2387

41. Siles JA, Thompson LQZ (2010) Biorefinery of waste orange peel. Crit Rev Biotechnol 30(1):63-69

42. Silva D, da Silva-Martin E, da Silva R, Gmes E (2002) Pectinase production by *Penicillium viridicatum* RFC3 by solid state fermentation using agricultural wastes and agroindustrial byproducts. Braz J Microbiol 33(4):1-10

43. Sittidilokratna C, Suthirawat S, Chitradon L, Punsuvon V, Vaithanomsat P, Siriacha P (2007) Screening for pectinase producing bacteria and their efficiency in biopulping of paper mulbery barks. Sci Asia 33:131-135

44. Sneath PH (ed) (1986) Bergey's Manual of systematic Bacteriology (Volume 2) Williams & Wilkins, Baltimore, London, Los -Angeles, Sydney.

45. Soares MMCN, Da Silva R, Gomes E (1999) Screening of bacterial strains for pectinolytic activity: characterization of the polygalacturonase produced by *Bacillus* sp. Rev Microbiol 30(4):299-303

46. Solís-Pereira S, Favela-Torres E, Viniegra-González G, Gutiérrez-Rojas M (1993) Effects of different carbon sources on the synthesis of pectinase by *Aspergillus niger* in submerged and solid state fermentations. Appl Microbiol Biotechnol 39(1):36-41

47. Souza JBV, Silva ES, Maia MLS, Teixeira MFS (2003) Screening of fungal strains for pectinolytic activity: endopolygalacturonase production by *Peacilomyces clavisporus* 2A.UMIDA.1. Process Biochem 39(4):455-458

48. Wilkins MR, Suryawati L, Maness NO, Chrz D (2007) Ethanol production by *Saccharomyces cerevisiae* and *Kluyveromyces marxianus* in the presence of orange-peel oil. World J Microbiol Biotechnol 23(8):1161-1168

Chapter 8

Proximate Composition, Mineral Content and Antinutritional Factors of Brebra (Millettia ferruginea) Seed Flour as well as Physicochemical Characterization of its Seed Oil

Berhanu Andualem and Amare Gessesse

[1]Department of Biotechnology, Natural and Computational Sciences Faculty, University of Gondar, P.O. Box 196, Gondar, Ethiopia

[2]Biotechnology Institute, Collage of Natural Sciences, Addis Ababa University, Addis Ababa, Ethiopia

ABSTRACT

Still there is no scientific report about the proximate analysis of seeds and characteristics of oil produced from brebra seed. Objective of this study was to determine proximate and antinutritional characteristics of seeds as well as the physicochemical characteristics of brebra seed oil. Crude oil, protein, fiber, ash, moisture and carbohydrate content of brebra were 48.5 ± 0.99%, 29.7 ± 0.23%, 2.41 ± 0.12%, 3.24 ± 0%, 4.24 ± 0.04% and11.92 ± 0.2%, respectively. Seed has concentrated energy (6.0298 Kcal/gm). The respective tannin, oxalate and phytic acid value were 84.3 ± 0.89 mg/100 gm, 20.97 ± 0.36 mg/100 gm and 291.62 ± 0.87 mg/100 gm, respectively. Cyanide was not detected in the sample. Seed contains high concentration of phosphorus (1062.1 ± 0.3 mg/100 g), potassium (281 ± 0.1 mg/100 g), magnesium (112.38 ± 0.1 mg/g), sodium (93.26 ± 0.1 mg/g) and calcium (61.55 ± 0.01 mg/g). The oil was analyzed for specific gravity at 20°C, viscosity at 40°C, refractive index at 40°C, acid value, saponification value, iodine value, peroxide value and ester value. Their respective values were 0.942, 40.59 mm^2/s, 1.473, 0.39 mg KOH/g, 174.95 mg KOH/g, 104.48 $gI_2/100$ g, 6.88 and 174.56 mg KOH/g. Unsaturated fatty acids accounts (80.7%), of which 48.2% and 27.7% were linolcic and linolenic, respectively, which make suitable for production of biodiesel. Seed has higher nutrient composition, low antinutritional elements and high calorie value compared to some legumes.

BACKGROUND

Millettia ferruginea (Hoechst.) Baker is a useful endemic tree species of Ethiopia with great potential for agroforestry. It is belonging to the family Fabaceae (Leguminosae) sub-family Papilionnodeae. This plant is known to have two subspecies, namely, *ferruginae* and *darassana* (Thulin 1983). Subspecies *ferruginae* is known to occur at North Ethiopia within the range of 1,000 and 2,500 m above sea level, while subspecies *darassana* is located in the southern part of the country (particularly in Sidamo) within the range of 1,600 and 2,500 m above sea level. The hybrid of the two subspecies believed to be found in the central and western part of Ethiopia (Thulin 1983). According to the flora of Ethiopia, vol. 3:108, (1989), *Millettia ferruginea* found in the following regions: *subspecies ferruginea; Tigray, Gondar, Gojam, Shewa, Welega and Hareg.* The subspecies*darassana* is commonly found in Welega, Shewa, Harege, Bale, Ilubabor, Kefa and Sidamo. Any way both species are believed to be found only in Ethiopia, despite as yet an unconfirmed report from the Sudan (Thulin 1989).

Millettia ferruginea is a N_2 fixing leguminous tree species that is known to have positive effects on associated crops in the southern parts of Ethiopia (Machachlan 2002). The tree usually occurs on farmlands in association with some important annual and perennial crops, such as barley (*Hordeum vulgare* L.), *Ensete ventricosum* Welw.) Cheeeman, maize (*Zea mays* L.), sorghum (*Sorgum bicolor* (L.) Moench s.l. and coffee (*Coffea arabica* L.) in the Wendo-Genet, Sugallae and Sokicha areas (Southern Ethiopia). According to Muleta (2007), from 14 dominant coffee shade tree species,*Millettia ferruginea* had the highest frequency of occurrence (22.3%). Soil near *Millettia ferruginea*tree is found to be rich in nutrient (Hailu et al. 2000). Its wood is used to fire wood, house construction, flowers serve as feed for bees; leaves, shoots and flowers are used as fodder for ruminants. The pulverized and crushed seeds widely used as fish poisons (Negash 2002) and as insecticides for scabies and chiggers (Stein 1973). It is also used for erosion control MacLachan (2002). Hailu et al. (2000), presented a reproduced summary of the use of *Millettia ferruginea* trees based on interviews made in Gedeo, Southern Ethiopia. Despite these significant benefits of the plant under this investigation, the plant seed protein and oil are not yet explored for production of economically important products like oil for soap and biodiesel production and oilcake for media and protein production for human and animal consumption. Currently, production of biodiesel from non-edible vegetable oil (like brebra oil) is considered important than that of using edible vegetable oil for biodiesel. Still there is no scientific report about the proximate analysis of seeds and characteristics of oil produced from brebra seed. Therefore, the main objective of this study was to determine proximate and antinutritional characteristics of seeds as well as the physicochemical characteristics of brebra seed oil. Such information may expand the scope of knowledge on the utilization and quality of the extracted oil and oilcake of the seed for different purposes.

RESULTS AND DISCUSSION

Sample Preparation

In this study, the processes of harvesting of seeds and oil extraction methods were developed and optimized. Brebra is a name given to *Millettia ferruginea* in Amharic. This Amharic name was literally inherited from behavior of the mechanism of seed dispersal nature, which is the seed mechanically dispersed about 20 meters in average far from the tree in explosive manner. This nature of seed dispersal mechanism poses a problem for seed harvesting. To overcome harvesting problem, the matured pale yellow pods were collected from the tree and covered with *teff* straw for certain period of time to accomplish its maturity. After maturation, the pods were put into a fiber sac to facilitate aeration and dried there in the sac and lastly released seeds collected in the sac. Fiber sac can provide free ventilation of air in order to avoid deterioration of seed quality by fungi. This method was originally adopted from the society. It is well known that *Millettia ferruginea* contains a chemical compound that is found to be toxic for fish is known as rotenone (Dagne et al. 1990), which is widely used by the society for fishing. The society was collected the seeds and pods of the tree for fishing by the method already mentioned above.Out of 100 kg of dry pods with seeds, 25 kg (25%) seed was harvested. From an average sized tree, about 150 kg of pods containing seeds can be harvested. From a single tree it is possible to produce 37.5 kg seeds. From one hectare of land in average it is possible to plant about 35 trees. Therefore, from one hectare land one can harvest 1350 Kg dry weight of seeds. The whole process of harvesting and extraction of oil is shown on Figure 1. Pure oil was extracted by the help of co-solvent amended (hexane and ethanol) techniques. Ethanol was employed to remove any polar residues from oil. Ethanol soluble phospholipids, proteins and other polar substances in the oil were moved to the ethanol phase and pure oil remains in the hexane phase. Both solvents were recovered from their perspective mixture by means of Rota vapor.

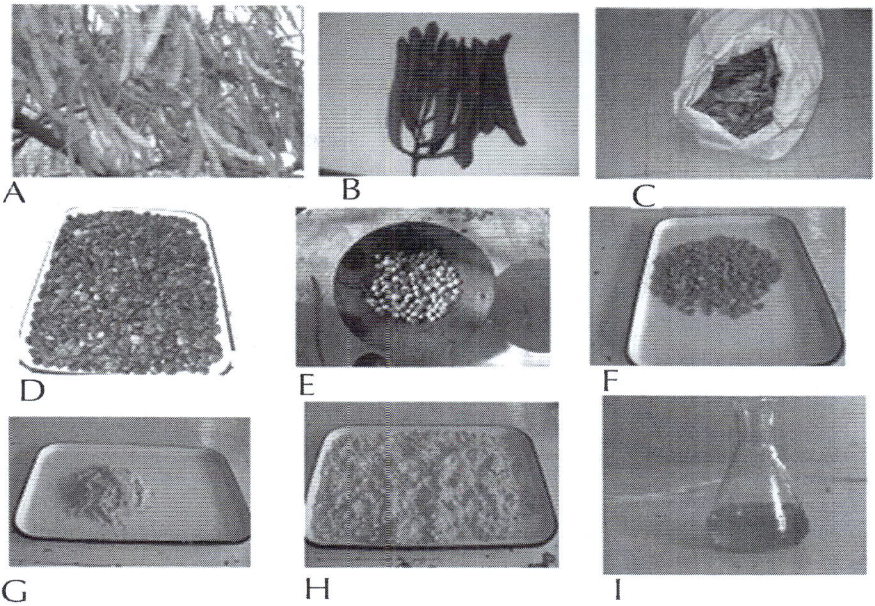

Figure 1: The process of harvesting of seeds and oil extraction from brebra tree. A = a typical brebra tree with pods, B = pod with seeds, C= pods in sac, D = harvested seeds, E = when seeds were dehulled by roasting with pan, F = dehulled seeds, G = powdered seeds prior to extraction of oil, H = defatted flour of brebra seed and I = extracted oil.

Proximate Composition

Table 1 presents the result of the proximate chemical composition (% dry weight) of brebra seed. The results reveal brebra seed as an oilseed with a potential of high oil and protein content to satisfy calorie and protein demand of the populations. The extracted brebra oil with the average of 48.5% is in close agreement with the average value of 49.5% melon oil seeds (Lge et al. 1984), 48.1% pumpkin seed (Fagbemi & Oshodi, 1991), 48.9% conophornut (Enujiugha 2003), 49.1% cashew nut (Akinhanmi, et al. 2008), 50% castor seed, 50% sesame seed, but is higher than*Crambe abyssinica* oil seed, 45.4 (Massoura et al. 1996), 42% groundnut kernel, 37% rapeseed, 36% palm kernel, 35% mustard, 32% sunflower, 20% palm fruit, 13% cotton seed and 23.5% soybean (Paul and Southgate 1980). The very high oil content suggests that brebra can be used as potential source of raw material for commercial activities. In brief, it can serve as feedstock for production of biodiesel, glycerol, soap and economically important materials, but not as nutrition at this level unless further investigation is carried to remove mild toxic substance, rotenone.

Table 1: Proximate chemical composition of non-defatted flour of brebra seed (gm/100 gm dry matter)

Component	Mean ± S.D (%)	CV (%)
Crude oil	48.5 ± 0.99	2.04
Crude protein	29.7 ± 0.23	0.77
Crude fiber	2.41 ± 0.12	4.98
Ash	3.24 ± 0.0	0.0
Moisture	4.24 ± 0.04	0.94
Carbohydrate (by difference)*	14.32 ± 0.2	1.41
Dry matter	95.8 ± 0.07	0.07
Organic matter	92.52 ± 0.03	0.03
Nitrogen free extract (NFE)**	11.91 ± 0.2	1.68
Energy (Kcal/g)***	6.0298 ± 0.01	0.17
Rotenone	0.701 ± 0.02	2.85

All tests were performed in triplicates and mean values are taken.

Carbohydrate = 100- (H2O + Ash + Cp + EE).

NFE = 100 - (H_2O + CP + CF + EE + Ash).

Energy (kcal) = 4 (g protein + g carbohydrate) + 9 (g lipid).

CV% = S.D/mean × 100.

Andualem and Gessesse *SpringerPlus* 2014 3:298 doi: 10.1186/2193-1801-3-298

The amount of crude protein in brebra seed was 29.7%, which was higher than protein rich foods such as quinoa (Ogungbenle et al. 2009), bambara groundnut (Yagoub and Abdalla 2007), cowpeas (Ragab et al.2004), seeds ranging between 13.5-26.8%. Moreover, chick beans, 19.4%, lima bean, 19.8% (FAO, 1982), *Crambe abyssinica*, 25.1% (Massoura et al. 1996), pea, 20.1%, (Sumner et al.1980) and cashew nut, 25.5% (Aremu et al. 2006), kidney beans, 20.9% and lentils, 22.9% (Perez-Hidalgo et al.1997) have less amount of proteins in comparison with that of brebra seed flour protein. On the other hand, the amount of brebra seed protein is almost equal to conophor nut, 29.1% (Enujiugha 2003), jack bean, 30.8% (Anonymous 1972), *Canaralia cathartica*, 31.2% (Seena and Sridhar 2006) and roselle (32.3%) (Mohammed et al. 2007) but less than lenti, 33.4% (Suliman et al.2006), *Cataralia maritime*, 34.1% (Seena and Sridhar 2006), ashew nut, 36.3% (Akinhanmi et al. 2008), soybeans, 37% (Messina 1997) and barbados, 48.1% (Yusuf et al. 2007). This high quantity of protein can serve as media for microorganisms, fed for animals and even can serve as human food after detailed investigation. The value obtained for carbohydrate (11.92%) in this study is incomparable with an acceptable range of values of legumes, 20-60% of dry weight (Arkroyed and Doughty 1964) but almost equal to carbohydrate content of conophor nut (Enujiugha 2003) and even greater than cashew nut (Akinhanmi et al. 2008). This result thus gave us indication that the energy source is largely oil and in some extent protein (through deamination).

The calculated metabolizable energy value (6.03/Kcal/g) is higher than 3.12 Kcal/gm in *C. palcherria*seed, 4.49 Kcal/gm in *G. affricanum* seeds (Ekop 2007) and 5.46 Kcal/gm in fish *Citharinun citharus* reported by (Abdullahi 1999) in Nigeria. The energy content of barbados (3.12 Kcal/gm) was smaller than brebra seed flour energy content Yusuf et al. (2007)). According to this finding, brebra seed has concentrated energy supply in comparison with the above reported energy sources.

The moisture value of the oil in this study which was 4.24% is somehow low when compared with the value of moisture of legumes ranging between 5.0% and 11% reported in the literatures (Aremu et al. 2006; Lge et al. 1984). Ash content of brebra seed, which is an indicator for mineral elements, in this study was 3.24%, which is closely comparable with ash values of 3.68%, 3.22% and 3.56% reported for pigeon pea, lima bean and lablab bean, respectively (Aletor and Aladetimi 1989). It has been recommended by Pomeranz and Clifto (1981) that ash contents of seeds and tubers should be in the range 1.5-3.5% in order to be suitable for animal feeds. In this case, the ash content of this study fall within this range hence it can be recommended for animal feeds and human consumption as well as it can serve as microbial media without mineral supplement.

The dry matter and organic matter content of brebra seed flour were 95.8% and 92.52%, respectively. These show that the seed flour contains high amount of organic matter. The amount of rotenone of the seed was 0.701 ± 0.02%. It can be used as insect pesticide if production and application method is developed.

Table 2 depicts the amino acid composition of brebra seed. Glutamic acid (18.62 g/100 g) was the most predominant amino acid followed by aspartic acid (6.43 g/100 g), leucine (2.98 g/100 g), and lysine (1.79 g/100 g). The values of amino acids showed that cysteine and methionine were in the lowest levels. On the other side, essential amino acids represented 11.88 g/100 g, while nonessential amino acids represented 32.12 g/100 g and E/N ratio was 0.37. Total amino acids of brebra (44.0 g/100 g) are greater than soybean total amino acids (36.62 g/100 g). However, total essential amino acids of soybean (13.52 g/100 g) are greater than brebra seed essential amino acids (11.88 g/100 g).

Table 2: Total amino acid composition of brebra seed (g/100 g)

Amino acids	Brebra seed	Soybean seed*
Essential amino acids		
Isoleucine	1.63	1.971
Leucine	2.98	3.309
Lysine	1.79	2.706
Methionine	0.12	0.547
Phenylalanine	1.75	2.122
Threonine	0.97	1.766
Histidine	0.70	1.097
Non-essential amino acids		
Alanine	1.33	1.915
Ornithine	0.02	-
Asparagine	1.60	-
Aspartate	6.43	5.112
Cysteine	0.09	-
Glutamate	18.62	7.874
Glycine	1.10	1.880
Proline	1.20	2.379
Serine	1.21	2.357

Tyrosine	0.52	1.539
Total free NEAA	0.04	-
Total free EAA	0.40	-
Total essential amino acids (E)	11.88	13.52
Total nonessential amino acids N)	32.12	23.10
E/N	0.37	0.59
Total AAS	44.00	36.62

Source: USDA Nutrient database, 2010

Andualem and Gessesse *SpringerPlus* 2014 3:298 doi: 10.1186/2193-1801-3-298

Amino acid profile of brebra seed protein was compared with the well-known amino acid composition of soybean. Total amino acid composition of brebra is far greater than the total amino acids of soybean (USDA Nutrient database 2010). It is rich in both essential and non-essential amino acids. As most edible legumes (Bhagya et al. 2007), lysine of brebra seed is higher than sulphur amino acids (cystine and methonine). Since brebra seed is rich in content of amino acids, it can use as a potential source of protein for human being.

Mineral Composition

Table 3 shows the mineral content of brebra seed. The abundant minerals were phosphorus (1062.1 ± 0.3 mg/100 g), potassium (281 ± 0.1 mg/100 g), magnesium (112.38 ± 0.1 mg/g), sodium (93.26 ± 0.1 mg/g) and calcium (61.55 ± 0.01 mg/g). All the mineral elements measured were found to be higher than conophor nut, cashew nut, and bean seeds (Table 3) (Akinhanmi ET al.2008; Aremu et al. 2006).

Table 3: Mineral composition of defatted brebra seed flour

Mineral	mg/100 g			
	Brebra seed	*Conophor nut	**Cashew nut Kerrel	***Ripened beans
Magnesium (Mg)	112.38 ± 0.1	57.37 ± 2.53	19.3 ± 0.1	28.7 ± 2.8

Calcium (Ca)	61.55±0.01	42.06±2.01	21.5±0.0	140.0±7.8
Sodium (Na)	93.26±0.1	-	8.2±0.2	60.8±5.0
Zinc (Zn)	2.0±0.2	6.84±0.02	0.8±0.1	10.7±0.7
Iron (Fe)	27.81±0	1.55±0.08	0.6±0.1	1.2±0.1
Potassium (K)	281.00±0.1	-	27.5±0.4	1327.0±2.3
Manganese (Mn)	25.5±0.2	-	-	2.02±0.1
Copper (Cu)	17.39±0.1	1.56±0.05	-	0.34±0.1
Phosphorus (P)	1062.1±0.3	465.95	14.0±0.2	214±14.1

Enujiugha, 2003 Akinhanmi ET al.2008, Bhagya et al. 2006.

Andualem and Gessesse *SpringerPlus* 2014 3:298 doi: 10.1186/2193-1801-3-298

In addition to its high protein content, brebra seed contains a high concentration of minerals, especially phosphorus, potassium, magnesium sodium and calcium. The mineral content of brebra seed is in general higher than those reported for other legumes (Akinhanmi et al. 2008; Aremu et al. 2006) as mentioned above. It has a potential to supply sufficient amount of minerals for consumers and microbial media for microorganisms.

Antinutritional Factors

Antinutritional components of brebra seed is summarized in Table 4. The presence of antinutritional factors in the sample is of significant importance since they are some deleterious effects on both humans and other animals, for instance, oxalate is a chelating agent, which binds calcium very effectively. Plants with high oxalate content may produce acute metabolic calcium deficiency (hypocalcemia) when we use plant product as a main food source (Checke and Shull, 1985). The concentration of oxalate (22.97 mg/100 gm) in the seed in this investigation seems to be on the low side when compared to reported values in some crop seeds (Umoren et al. 2005).

Table 4: The antinutritional factors of brebra seed flour

Component	Mean ± S.D (mg/100 gm)
Tannin	84.3 ± 0.89
Cyanide	B.D.L
Oxalate	20.97 ± 0.36
Phytic acid	291.62 ± 0.87

B.D.L = below detection limit.

Andualem and Gessesse *SpringerPlus* 2014 **3**:298 doi: 10.1186/2193-1801-3-298

Tannin is known to evoke growth-depressing effects in rats. In this study, the tannin level (84.3 mg/100 gm) was found to be relatively high in comparison with tannic acid found in some literatures (Akinyede et al. 2005; Enujiugha 2003; Umoren et al. 2005). However, the tannin content of this study is less than the tannin content of some dry bean seed varieties (930 mg/100 gm) (Deshpande et al. 1986). The tannin amount in this study may not be as such harmful as expected for consumption. High amount of tannins are well known to form complex with proteins and reduced the solubility of proteins and make protein less susceptible to proteolytic attack than the same proteins alone (Carbonaro et al. 1996). However, relatively some amount of tannin, like this finding, may have a potential role as protective factors against free radical mediated pathologies, such as cancer and atherosclerosis, in humans (Kehrer 1993). According to Bawadi et al. (2005) report, water-soluble condensed tannins extracted from black beans inhibited the growth of MCF-7, Caco-2 colon, and Hs578T breast as well as DU 145 prostatic cancer cells. Other findings, associating polyphenols (including tannins) to free radicals scavenging and metal chelating activities, suggested their potentially beneficial implications in the treatment and prevention of cancer (Hangen and Bennin 2002).

Cyanogenic glycoside contents of legume seeds have been investigated. According to Liener (1977) report, total cyanide values for different legumes such as cowpea, lima bean, field pea, kidney bean, chicken pea and pigeon pea, were 2.1, 210–312, 2.3, 2.0, 0.8 and 0.50 mg/100 gm, respectively. About 0.40 mg/100 gm of cyanogenic glycoside from *Milletia obanensis* was also reported by Umoren et al. (2005). In this study, the cyanide content was found to be below detectable level. In terms of cyanide content, brebra seed flour is safe to use as food for both humans and other animals. In summary, the antinutritional content of brebra seed flour is not out of the range value of different legumes seeds and other crops reported by other literatures.

Physicochemical characterization of oil

The physicochemical properties of brebra are shown on Table 5. Oil is clearer brown yellow in colour and less viscous, 40.59 CTm at 40°C than cold break seed oil of 57.5 CTm at the same temperature (Catarelli et al. 1993). It has density and specific gravity of 0.942 and 0.926 at 20°C, respectively. The specific gravity of this study is lower than the specific gravity of cashew nut (0.964) (Aremu et al., 2006) and caster seed oil (0.958) (Akpan et al. 2007). The refractive index of this study (1.473) is in close agreement with 1.465 (Aremu et al. 2006), 1.462 (Akintayo and Bayer 2002) and 1.468 (Akpan et al. 2007) of castor, akee pulp and cashew nut seed oils, respectively. This implied that brebra oil is less thicker than most of drying oils whose refractive indices range from 1.475 to 1.485 (Duel and Tr 1951).

Table 5: Brebra oil chemical and physical characteristics

Oil physicochemical characteristics	Unit	Values
Colour		Pale-yellow color limpid liquid
Density at 20°C	Kg/m3	0.942
Specific gravity at 20°C	kg/l	0.926
Kinematic Viscosity at 40°C	mm2/s	40.59
Acid value	mg KOH/g	0.39
pH value		6.38
Saponification value	mg KOH/g	174.95
Refractive index at 40°C		1.473
Iodine value	gI2/100 gm	104.48
Peroxide value	mEq/Kg	6.88

Ester value		174.56

All values are mean value of triplicates.

Andualem and Gessesse *SpringerPlus* 2014 3:298 doi: 10.1186/2193-1801-3-298

The saponification value of the oil in this study was 174.95 mg KOH/gm. This was lower than the values for some common oils like castor seed oil (185.83) (Akpan et al. 2007), palm oil (196-205 mgKOH/g), groundnut oil (188-196 mgKOH/g), corn oil (187-196 mgKOH/g) as reported by Akinhanmi et al. (2008), coconut oil (253 mg KOH/gm) and palm kernel oil (247 mg KOH/gm) (Pearson, 1976). However, this saponification value is within the same range of some edible oils reported by Eromosele and Paschal (2002). Moreover, saponification value (174.95) of the oil in this investigation is almost within the range of (175–187) ASTM (2002) specification for oils. According to Pearson (1976), oils with lower saponification values contain high amount of long chain fatty acids. Therefore, the value obtained for brebra seed oil contained high quantity of higher fatty acids (as it is presented on Table 6, fatty acids ≥ 18 carbon chain accounts 92.8%). The value of viscosity of brebra oil was 40.49 mm^2/s (as mentioned above) is found to be less than the viscosity of cashew nut oil (56 mm^2/s) (Akinhanmi et al. 2008). For production of biodiesel, therefore, transesterification reaction is the best method to reduce the viscosity of the oil. The oil under investigation has very low acid value of 0.39 mg KOH/g when compared with cashew nut oil (0. 82 mg KOH/g) (Aremu et al. 2006), refined castor oil (0.869) and crude castor oil (1.148) (Akpan et al. 2007), plukenetia conophoora (11.5 mg KOH/g) as reported by Akintayo and Bayer (2002), aenniseed (47.6%) by Ohsodi (1992). The acid value in this oil is below the maximum limit (2.0 mg KOH/g) of DIN EN ISO 660 and nearly within the range of ASTM specification (0.4 - 4.0) of castor oil (ASTM 2002).

Table 6: Fatty acid composition of fatty acid of brebra oil (*FAs = fatty acids)

No	Fatty acids	Carbon number	Quantity (%)
	Unsaturated FAs*		
1	Oleic	C18:1	0.6
2	Linoleic	C18:2	48.2
3	Linolenic	C18:3	27.7

4	Arachidonic	C20:4	0.8
5	Eurcic	C22:1	3.4
	Total		80.7
	Saturated FAs*		
6	Palmitic	C16:0	7.2
7	Stearic	C18:0	1.9
8	Arachidic	C20:0	3.3
9	Behenic	C22:0	6.9
	Total		19.3

Andualem and Gessesse *SpringerPlus* 2014 **3**:298 doi: 10.1186/2193-1801-3-298

The low saponification value and acid value imply that the oil is more appropriate for biodiesel since the oil acid value for biodiesel, which is less than 1%. The iodine value of brebra oil was 104.48 gI_2/100 gm (within the range, 100-120 gI_2/100 gm, of DIN) which is greater than the range of 77–94 gI_2/100 gm olive oil, 8–10 gI_2/100 gm coconut oil, 12–18 gI_2/100 gm palm kernel, 44–58 gI_2/100 gm palm oil, 85–95 palme oleine, 20–45 gI_2/100 gm palme stearine, 50–60 gI_2/100 gm tallow, 60–70 gI_2/100 gm lard, (http://dec2.tec.agrar.tu-muenchen.de/pflanzoel/rkstandarde.html*webcite*) 44.4 gI_2/100 gm cashew nut oil (Aremu et al. 2006), 38.1 gI_2/100 gm citrullus vulgaris (Achinewhu 1990), *Hausa melon* seed oil (38.50 gI_2/100 gm) (Oladimeji et al. 2001) and 84.8 gI_2/100 gm refined castor oil (Akpan et al. 2007). One the other hand, the iodine value of this study was lower than the range of 110–115, 125–135, 125–140, and 115–124 gI_2/100 gm value of rapeseed oil, sunflower oil, soybean oil and corn oil, respectively. The oil in this study is considered drying oil since drying oils have an iodine value above 100 gI_2/100 gm (Duel and Tr 1951). The peroxide value of brebra oil was 6.88 mEq/Kg. It was greater than the peroxide value of 3.1 mEq/Kg of cashew nut oil (Aremu et al. 2006).

The fatty acid composition of brebra oil was determined by gas chromatography (Figure 2 and Table 6). The total unsaturated fatty acid content (80.7%) was higher than the total saturated fatty acids (19.3%) of brebra oil. Among polyunsaturated fatty acids linolcic (48.2%) and linolenic (27.7%) accounts 75.9% of the total fatty acids. The total unsaturated fatty acids (80.7%) of this study is greater than unsaturated fatty acids of 70.6% and 68.6% of cold break seed oil and hot break seed oil of tomato, respectively (Catarelli et al. 1993). Palmitic (7.2%)

and behenic (6.9%) acids are the major saturated fatty acids in the oil under investigation. In summary, there is high degree of unsaturation with long chain fatty acids.

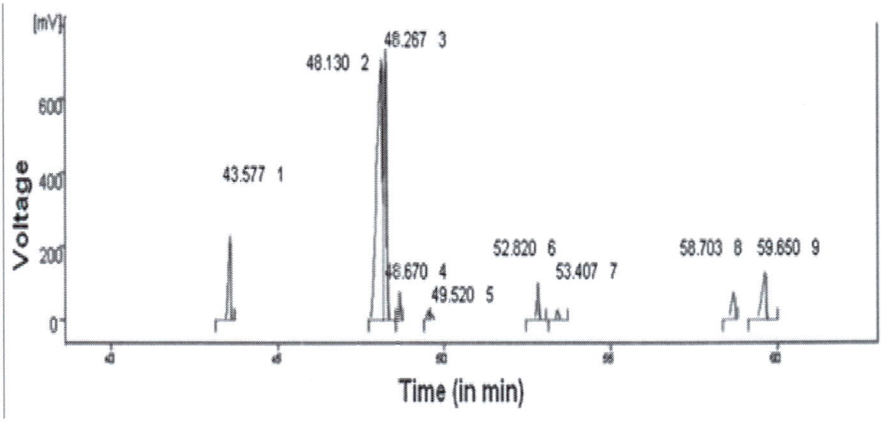

Figure 2: Chromatogram results obtained from brebra oil.

CONCLUSIONS

From the result of the analysis, it can be shown that the seed flour of brebra has higher nutrient composition and calorie value compared to some legumes most especially in terms of crude oil and protein. The amount of protein in the seed is high in comparison with that of most protein rich crops. However, it may not serve as a good source for carbohydrates compared to other food sources. With regard to antinutritional studies, the antinutritional content of the sample under investigation is not out of the range value of different crops reported by other literatures. The percentage oil content of brebra seed was found to be 48.5%. The oil produced in this study was analyzed for specific gravity at 20°C, viscosity at 40°C, refractive index at 40°C, acid value, saponification value, iodine value, peroxide value and ester value. Their respective values are 0.942, 40.59 mm^2/s, 1.473, 0.39 mg KOH/g, 174.95 mg KOH/g, 104.48 $gI_2/100$ g, 6.88 and 174.56 mg KOH/g. The oil is also composed of 80. 7% unsaturated and 18.3%% saturated fatty acids. The physicochemical properties of the oil indicated that it is drying oil, which is rich in content of unsaturated fatty acids. Fatty acid composition of brebra seed oil make suitable for production of biodiesel and soap. These characteristics of oil make suitable for production of biodiesel and soap. Thus, oil is of good quality and could be recommended as suitable for industrial usage. To our knowledge, this is the first scientific report about the production and characterization of protein and oil from brebra tree, which is endemic in Ethiopia, by using standard oil test methods and standard parameters.

MATERIALS AND METHODS

Harvesting and Sample Collection

Harvesting process was adopted from traditional method of the society. Matured (pale yellow colored) pods of brebra from the study plant were collected and covered with the straw of teff (*Eragrotis teff*) for more than a week and then collected in the fiber sac, which is used to ventilate in order to avoid spoilage by fungi. The matured seeds were selected in order to improve the oil meal quality and to increase the capacity and efficiency of the extraction plant. The seeds were dried by using oven at 60°C more than 8 hr. The moisture content of the seed was determined by heating at 110°C for 24 hr in an oven by the procedure described by AOAC (1990). The seed coat of the seeds was dehulled by lightly roasted on pan and in the process water was added to sequester the seed coat and lastly dehulled by wooden mortar and pestle. For oil extraction, solvent (hexane) treatment techniques was used. To refine the oil co-solvent system technique (hexane and ethanol) was used. The process of refinery of the oil was determined and optimized in our previous study (Andualem and Gessesse 2012).

Proximate Analysis of Seed

The methods used for sample treatment and analysis were carried out based on the standard procedures recommended by AOAC (1990). Crude fat, ash, total carbohydrates, total nitrogen and nitrogen free extract were determined according to AOAC (1990). Oil extraction was carried out by using hexane as a solvent. Brebra seeds were ground with blender (Waring blendor) and the fine flour was mixed with hexane and the whole content was stirred by magnetic stirrer for more than 4 hr and then filtered with Whatman's No 1 filter paper. Hexane was recovered by the help of Rota vapor (Buchi, Switzerland) (Meher ET al.2006) at 100 rpm. Total oil was quantified gravimetrically and calculated as percentage of oil. Protein (N × 6.25) was determined by the Kjeldahl method. To determine the ash content of the sample, 5 gm of the sample was incinerated in a muffle furnace. Crude fiber content of the sample was determined by mixing of the fine powder of the sample with 1.25% sulfuric acid and 1.25% sodium hydroxide solutions under specific conditions for ignition and dried residue remaining after digestion of the samples was considered as crude fiber (AOAC 1990). Calories were calculated by multiplying the amount of protein, carbohydrate and fat by the factors of 4, 4 and 9 (K cal) and 17, 17 and 37 (KJ), respectively, (EEC, 1990). To determine the moisture, the sample was dried to a constant weight in a vacuum oven at 100°C (AOAC, 1990). The moisture loss was determined gravimetrically.

Determination of Amino Acid Composition

Materials and Reagents

The EZFaast GC-MS physiological amino acid analysis kit, Methanol (HPLC grade) and the internal standard and additional amino acid standards were obtained from Phenomenex (Cheshire, UK), (VWR, Leicestershire, UK) and Sigma (Dorset, UK), respectively.

Sample Extraction (Five Replicates per Treatment)

To determine total amino acid, 15 ± 0.03 mg sample was mixed to 1.00 ml 6 N HCl in 2 ml screw cap vial. The caps were taped in place with autoclave tape and heated in an oven at 110°C, 24 h. On removal from the oven, samples were cooled and 100 µL 0.75 mM norvaline solution added. Samples were mixed thoroughly and evaporated in a centrifugal vacuum concentrator (ThermoSavant). The residue was reconstituted in 1 ml distilled water and filtered through a 0.2 µM PTFE syringe filter (Pall Acrodisc) into a fresh vial. A 50 µl portion was transferred to a reaction tube (supplied with the EZFaast kit) and the solvent removed in a centrifugal sample concentrator (ThermoSavant).

In free amino acid analysis, 15 ± 0.03 mg sample was transferred to a 1.5 ml microfuge tube and 1 ml, 0.075 mM norvaline solution in 80:20 H_2O: MeOH added. Samples were heated at 50°C, 10 min, and cooled for 2 min and centrifuged at 13400 rpm for 10 min. Lastly, 750 µL supernatant was transferred to a reaction tube (supplied with kit) and the solvent removed in a centrifugal sample concentrator (ThermoSavant). Sample residues were stored, desiccated at -18°C.

Samples were reconstituted in 0.2 ml 80:20 H_2O: MeOH with vortexing to ensure complete dissolution. Amino acids were isolated from the samples by ion exchange and derivatized to their propyl-chloroformate derivatives according to the protocol supplied with the EZFaast'.

GC-MS Analysis

Samples were analysed on a Hewlett Packard 5975C Inert MSD coupled to a 7890A Gas Chromatograph fitted with a Zebron Amino acid ZB-AAA column, split/splitless injector and MPS2 automatic liquid sampler. Two µl splitless injections were adjusted at purge time of 2 min with purge flow rate of 20 ml/min and an injector temperature was maintained at 220°C. The oven was programmed from 75°C (2.0 min) to 320°C (1.83 min) at 30°C/min was used with helium as the carrier gas at 9.5 kPa (1.4 ml/min), constant pressure. The source, quadrupole and transfer line temperatures were set at 230°C, 150°C and 320°C, respectively. Mass spectra were acquired at 70 eV over 45–450 m/z from 3–12 min with an acquisition rate of 3.5 Hz.

Data Analysis of Amino Acids

Data were quantified on the basis of extracted ion chromatograms (EIC) using the QuanLynx module of MassLynx 4.0 (Waters, Manchester, UK). The results were exported to Microsoft Excel (2003) and sample means and 95% confidence intervals (n = 5) were calculated for the free and total amino acid composition of the flour sample. Calibration curves from 0–26667 $pmol.mg^{-1}$ F.W. and 0–2667 $nmol.mg^{-1}$ F.W. (for the total and free amino acid respectively) were prepared and analysed alongside the samples.

Mineral Composition

To determine the mineral content of defatted flour, 5.0 g sample was incinerated in a furnace at 500°C and the residues dissolved in 50 ml of 2.5% HNO_3 solution. The concentrations of Na, Ca, Mg, Fe, P, K, Zn, Cu, Fe, and Mn was determined using atomic spectrophotometer (Buck Science) absorption, following the method of Angelucci and Mantovani (1986). A calibration curve was prepared using standard metal solutions. Phosphorus was determined using the ammonium molybdate/ammonium vandate method of Chapman and Pratt (1968).

Anti-Nutritional Factors of Seed Flour

Estimation of Tannins

Tannins were estimated by Vanillin-HCl method of Price et al. (1978). Five gram of defatted brebra seed flour was treated with acidic methanol for extraction of tannins. From the diluted extract, 1 ml of was mixed with 5 ml of freshly prepared vanillin-HCl reagent and the optical density was determined at 500 nm by using spectrophotometer. As positive control, catechin standards were used side by side with the sample. The results were expressed as mg/100 gm dry wt.

Determination of Phytic Acid

Phytic acid composition was analyzed according to Wheeler and Ferrel (1971) by using 2.0 gm of dehydrated sample. A standard curve was constructed and expressed the results as Fe $(NO_3)3$ equivalent. The amount of phytate phosphorus content was calculated from the standard curve by assuming that 4:6 iron to phosphorus molar ratio.

Determination of Oxalate

To determine oxalate in brebra seed flour, the samples were separated into two fractions using the following procedure: two grams of finely grounded brebra seed flour was extracted with 100 ml of boiling distilled water for 30 min, filtered and adjusted to 200 ml. On the other hand, the hot water extract residue was further extracted with 150 ml of boiling 1 M HCl for 30 min, adjusted to 200 ml and filtered. The two filtrates were combined together. The content of oxalate in the two fractions was analyzed based on the method of AOAC (1990) with the help of potassium permanganate titration. All the analyses were tested triplicate and the results calculated and expressed on dry weight basis.

Determination of Cyanide in Brebra Seed Flour

The content of cyanide in brebra seed flour was determined by the amount of HCN released on hydrolysis. Brebra seed flour extract was obtained by homogenizing 30 gm of flour in 259 ml of 0.1 M orthophosphoric acid for 5 min. The homogenate was centrifuged at 2,500 rpm for 20 min and clear supernatant was taken. An aliquot of the supernatant was used for determination of hydrogen cyanide using an auto analyzer Technicon AAII, according to the method of Rao and Hahan (1984).

Chemical Characterization of Brebra (Meillettia Ferrugeniea) Oil

After extraction of the oil, it was filtered to remove non-oil materials. A layer of sodium sulfate crystals was added to a flask and crude oil was added to remove any trace water. The dry agent was separated by decanting and filtration. The physicochemical determination of the oil for iodine value, saponification value and peroxide value were carried out according to the methods of AOAC (1990). Acid value was determined according to ASTM (2002). All tests were performed in triplicate.

Fatty Acid Analyses

The fatty acid profile was determined as fatty acid esters by gas chromatography. The sample methyl esters were prepared using the method used by the IOOC (International Olive Oil Council2001). Standards of the methyl esters of the fatty acids were lauric (C12), myristic (C14), palmitic (C16), stearic (C18: 0), oleic (C18:1), linoleic (C18:2), linolenic (C18:3), arachidic (C20:0), arachidonic (C20:4), behenic (C22:0), eurcic (C22:1) and lignoceric (C24:0). Standard mixtures of these esters were injected in the Gas Chromatography (DANI GC 1000) for identification and quantification purposes. Standards and samples (3 drops dissolved in 3 ml of chloroform) were injected (0.5 ml) to the GC. The column and mobile phase were ECT-5 and 5% phenyl and 95% methylpolysiloxane, respectively. The flow rate and pressure used were ml per minute and 1.25 Bar, respectively. The GC oven was kept at 50°C for 2 min, heated at 4°C/min up to 250°C, where it was kept for 15 min. The detector was a flame ionization detector (FID), and the carrier gas was nitrogen (5 ml/min) (Alcantara et al. 2000).

The content of the sample was quantified by comparing the FID counts for each methyl ester of the GC sample of methyl ester with the FID counts of each methyl ester in the standard mixture of fatty acid methyl esters (FAME's), averaging out these relationships for all the methyl esters (Alcantara et al. 2000).

Determination of Ester Value

The ester value is a measure of the amount of ester present in the given oil. It is expressed in the same terms as saponification value and the acid value. It was determined by subtracting the acid value from the saponification value (Ester value = Saponification value – Acid value).

Physical Characteristics of Oil

Determination of physical characteristics such as moisture, specific gravity and density, Kinematic viscosity, refractive index and pH value were carried out according to the methods of ASTM (2002).

AUTHORS' CONTRIBUTIONS

B-A Write and edits the whole manuscript and B-A supervised the overall activities and reviewed the documents. B-An also involved in statistical analysis and preparation of this manuscript. Both authors have read and approved the final manuscript.

ACKNOWLEDGEMENTS

We are thankful to the University of Gondar for financial assistance (UoG/Budget/no.6215) for the successful completion of this research work. We would like to express our appreciation to the Department of Biotechnology for the facilities provided during this investigation.

REFERENCES

1. Abdullahi SA (1999) Nutrient content of Citharinus citharus and C. latus. Zuma J Pure App Sci 2:66-68
2. Achinewhu SC (1990) Composition and food potential of melon seed (*C. vulgaris*). Nig Food J 8:130-133
3. Akinhanmi TF, Atasie VN, Akintokun PO (2008) Chemical Composition and Physicochemical Properties of Cashew nut (*Anacardium occidentale*) Oil and Cashew nut Shell Liquid. J Agri Food Enviro Sci 2(1): ISSN 1934-7235
4. Akintayo ET, Bayer E (2002) Characterization and some possible uses of Plukenetia conophora and Adenopus breviflorus seeds and seed oils. Biores Tech 85:95-97
5. Akinyede AI, Amoo IA, Eleyinmi AF (2005) Chemical and functional properties of full fat and defatted Dioclea reflexa seed flours. J Food Agri Envir 3(2):112-115
6. Akpan UG, Jimol A, Mohammed (2007) Extraction, characterization and modification of castor seed oil. Leonardo J Sci Issue 8:43-52 ISSSN 1583–02333
7. Alcantara R, Amores J, Canoira L, Fidalgo E, Franco MJ, Navarro A (2000) Catalytic production of biodiesel from soy-bean oil, used frying oil and tallow. Biomed and Bioen 18:515-527
8. Aletor VA, Aladetimi OO (1989) Compositional evaluation of some cowpea varieties and some under-utilized edible legumes in Nigeria. Die Nahrung 33(10):999-1007
9. Andualem B, Gessesse A (2012) Methods for refining of brebra *(Millettia ferruginea)* Oil for the production of biodiesel. World App Sci J 17(3):407-413

10. Angelucci E, Mantovani DMB (1986) Minerais em alimentos. SBCTA, ITAL, 13lp, Italy.
11. Anonymous (1972) Tropical Legumes: Resources Future. National Academy of Science Washington, DC, Washington. p 24
12. AOAC (1990) Official Methods of Analysis. Association of Official Analytical Chemists, Washington DC.
13. Aremu MO, Olonisakin O, Bako DA, Madu PC (2006) Compositional studies and physicochemical characteristics of cashew nut (Anarcadium occidentale) flour. Pakistan J Nutr 5(4):328-333
14. Arkroyed WR, Doughty J (1964) Legume in human nutrition. Food and Agricultural Organization nutrition studies publication. p 19
15. ASTM (2002) Petroleum Products, Lubricants, and Fossil Fuels. In: Annual Book of ASTM Standards. West Conshohocken, PA. USA: ASTM nternational.
16. Bawadi HA, Bansode RR, Trappey A, Truaxc RE, Losso JN (2005) Inhibition of Caco-2 colon, MCF-7 and Hs578T breast, and DU 145 prostatic cancer cell proliferation by water-soluble black bean condensed tannins. Cancer Letter 218:153-162
17. Bhagya B, Sridhar KR, Seena S, Bhat R (2007) Nutritional qualities of ripened beans of mangrove wild legume *Canavalia cathartica* Thouars. J Agrid Tech 3:255-274
18. Carbonaro M, Virgili F, Carnovale E (1996) Evidence for protein-tannin interaction in legumes: Implications in the antioxidant properties of faba bean tannins. Lebensm-Wiss Tech 29:743-750
19. Catarelli PR, Regitano-d'Arce MAB, Palma ER (1993) Physichochemical characteristics and fatty acid composition of tomato seed oils from processing wastes. Sci Agric (Piracicaba, Braz) 50:1Piracicaba feb./may 1993
20. Chapman HD, Pratt FP (1968) Ammonium Molybdate-Ammonium Vandate Method for Determination of phosphorus: Methods of Analysis for Soils, Plants and Water. California Univ. Public. Division of Agriculture and Science, phosphorus. pp 169-170
21. Checke PR, Shull LR (1985) Natural Toxicants in Feeds and Poisonous Plants. AVI Publishing Co., Inc., Westport, CT.

22. Dagne E, Bekele A, Noguchi H, Shibuya M, Sankawa U (1990) O-Geranylated and O-Prenylated flavonoids from *Millettia ferruginea*. Phytochemistry 29(8):2671-2673
23. Deshpande SS, Cheryan M, Salunkhe DK (1986) Tannin analysis of food products. CRC Crit. Rev Food Sci Nutr 24:401-449
24. Duel HJ, Tr (1951) The Lipids: Their Chemistry and Biochemistry Vol. 1. New York Inter Science Publishers, New York. pp 53-57
25. EEC (1990) Council directive on nutrition labeling for foodstuffs (90/496). Off J Eur Communities L 276:40-44
26. Ekop AS (2007) Determination of chemical composition of *Gnetum africanum* (Afang) seeds. Pakistan Jf Nutr 6:37-40
27. Enujiugha VN (2003) Chemical and functional characteristics of conophor Nut. Pakistan J Nutr 2(6):335-338
28. Eromosele CO, Paschal NH (2002) Characterization and viscosity parameters of seed oil from wild plants. Biores Tech 86:203-205
29. Fagbemi TN, Oshodi AA (1991) Chemical composition and functional properties of full fat fluted pumpkin seed flour. Nigerian Food J 9:26-32
30. FAO (1982) Food Composition Table for use in Africa Development of Health Education and Welfare. Rome Health Science and FAO Nutrition Division, Rome. p 32
31. Hailu T, Negash L, Olsson M (2000) Millettia ferruginea from southern Ethiopia: Impacts on soil fertility and growth of maze. Agrofor Syst 48:9-24
32. Hangen L, Bennin MR (2002) Consumption of black beans and navy beans (*Phaseolus vulgaris)* reduced azoxymethaneinduced colon cancer in rats. Nutr Canc 44:60-65
33. International Olive oil Council (2001) International Olive oil Council, Madrid.(2001). COI/T20/Doc. No.24
34. Kehrer JP (1993) Free radicals as mediators of tissue injury and disease. Crit Rev Toxic 23:21-48
35. Lge MN, Ogunsua O, Okon OL (1984) Functional properties of the protein of some Nigeria oil seeds.*Casophor* seeds and three varieties of some Nigeria oil seeds. Food Chem 32:822-825
36. Liener IE (1977) Removal of Naturally Occurring Toxicants through Enzymatic Processing. In: Finley RE, Whitaker JR

(eds) Advances in Chemistry Series no. 160, Food Proteins: Improvement Through Chemical and Enzymatic Modification, Washington, DC: American Chemical Society.

37. MacLachan M (2002) Manual of High land Ethiopian Trees (English/Amharic version). SIM Forestry Study project. Benawee Printing Press, Addis Ababa.

38. Massoura E, Vereijken JM, Kolster P, Derksen JTP (1996) Isolation and Functional Properties of Proteins from *Crambe Abyssinica* oil Seeds. In: Janick J (ed) Progress in new Crops, Alexandria, VA: ASHS Press. pp 322-327

39. Meher LC, Vidya SS, Dharmagadda SN, Naik (2006) Optimization of alkali-catalyzed transesteriWcation of Pongamia pinnata oil for production of biodiesel. Biores Tech 97:1392-1397

40. Messina M (1997) Soyfoods: Their Role in Disease Prevention and Treatment. In: Liu K (ed) Soybeans: Chemistry, Technology, and Utilization, New York, NY: Chapman & Hall. pp 443-477

41. Mohammed MA, Abu El Gasim A, Yagoub AE (2007) Furundu, from fermented/sprouted Roselle (*Hibiscus sabdariffa* L.) seed: investigation on chemical composition, antinutritional factors, HCl-extractability of minerals, amino acids composition, *In vitro* protein digestibility and microbial growth. Res J Agrie Biol Sci 3(6):876-885

42. Muleta D (2007) Microbial Inputs in Coffee (Coffea Arabica L.) Production Systems, Southwestern Ethiopia Doctoral Vol. 117. Diss. Dept. of Microbiology, SW.Actha Universitatis Agricuturae Sueciae, Sweden.

43. Negash L (2002) Research advances in some selected African trees with special reference to Ethiopia. Ethiop J Biol Sci 1(1):81-126

44. Ogungbenle HN, Oshodi AA, Oladimeji MO (2009) the proximate and effect of salt applications on some functional properties of quinoa (*chenopodium quinoa)* flour. Pakistan J Nutr 8(1):49-52

45. Ohsodi AA (1992) proximate composition, nutritionally valuable minerals and functional properties of adenopus brevriflorus bent seed flour. Food Chem 45:79-83

46. Oladimeji MO, Adebayo AO, Adgbesan AH (2001) Physico-chemical properties of *Hausa melon* seed flour. Ultra Sci 13:374-377

47. Paul AA, Southgate BAT (1980) Mucance and Widdowson's Composition of Food. Hermajesty's stationary office, London, UK. pp 227-228

48. Pearson D (1976) Chemical Analysis of Foods. Church Hill Livingstone, London.

49. Perez-Hidalgo M, Guerra-Hernandez E, Garcia-Villanova B (1997) Determination of insoluble dietary fiber compounds: cellulose, hemicellulose and lignin in legumes. Ars Pharma 38(4):357-364

50. Pomeranz Y, Clifto D (1981) Food Analysis Theory and Practices. In: Melon EE (ed) Chemical Food Analysis Practical Manual – Chemistry, Westport, C.T: Avi Publishing Company.

51. Price ML, Van Socoyoc S, Butter LG (1978) a critical evaluation of the vanillin reaction as an assay for tannin in sorghumgrain. Agri Food Chem 26:1214-1218

52. Ragab DM, Babiker EE, El Tinay AH (2004) Fractionation, solubility and functional properties of cowpea (*Vigna unguiculata*) proteins as affected by pH and/or salt concentration. Food Chem 84:207-212

53. Rao PV, Hahan SK (1984) an automated enzyme assay for determining the cyanide content of cassava (Manihot esculenta Crantz) and cassava products. J Sci Food Agri 35:426-436

54. Seena S, Sridhar KR (2006) Nutritional and microbiological features of little known legumes. Canavalia cathartica Thouars and C. maritima Thouars of the southwest coast of India. Current Sci 90(12):1638-1650

55. Stein J (1973) the Random House Dictionary of the English Language. Random House, New York.

56. Suliman MA, El Tinay AH, Elkhalifa AEO, Babiker EE, Elkhalil EAI (2006) Solubility as Influenced by pH and NaCl Concentration and Functional Properties of Lentil Proteins Isolate. Pakistan J Nutr 5(6):589-593

57. Sumner AK, Nielsen MA, Youngs CG (1980) Production and Evaluation of Pea Protein Isolate. Presented at the 40th Annual Meeting of the Institute of Food Technologist, New Orleans, La.June 8–11

58. Thulin M (1983) Leguminosae of Ethiopia. Opera Botanica 68:71

59. Thulin M (1989) Fabaceae (Leguminosae). In: Hedberg I, Edwards S (eds) Flora of Ethiopia, Volume 3, Pittosporaceae to Aralliaceae, Addis Ababa: Addis Ababa (Ethiopia) and Uppsala (Sweden). pp 49-251

60. Umoren UE, Essien AI, Ukorebi BA, Essien EB (2005) Chemical evaluation of the seeds of Milletia obanensis. Food Chem 91:195-201

61. USDA Nutrient database (2010) Search the USDA National Nutrient Database for Standard Reference. Department of Agriculture, Research Service. USDA National Nutrient, Beltsville, Maryland.

62. Wheeler EI, Ferrel RE (1971) Methods for phytic acid determination in wheat and wheat fractions. Cereal Chem 48:312-320

63. Yagoub AA, Abdalla AA (2007) Effect of domestic processing methods on chemical composition, *in vitro* digestibility of protein and starch and functional properties of bambara groundnut (*voandzeia subterranea*) seed. Res J Agri Biol Sci 3(1):24-34

64. Yusuf AA, Mofio BM, Ahmed AB (2007) Nutrient contents of pride of Barbados (Caesalpinia pulcherrima Linn) seeds. Pakistan J Nutr 6(2):117-121

Citations

CHAPTER 1

Lawrence J. B. Orikiriza, Hillary Agaba, Gerald Eilu, John D. Kabasa, Martin Worbes, and Aloys Hüttermann, Effects of Hydrogels on Tree Seedling Performance in Temperate Soils before and after Water Stress, doi:10.4236/jep.2013.47082.

CHAPTER 2

Y. El-Nahhal, M. Abadsa, and S. Affifi, "Adsorption of Diuron and Linuron in Gaza Soils," American Journal of Analytical Chemistry, Vol. 4 No. 7A, 2013, pp. 94-99. doi: 10.4236/ajac.2013.47A013.

CHAPTER 3

Chemura, A. , Kutywayo, D. , Chagwesha, T. and Chidoko, P. (2014) An Assessment of Irrigation Water Quality and Selected Soil Parameters at Mutema Irrigation Scheme, Zimbabwe. Journal of Water Resource and Protection, 6, 132-140. doi: 10.4236/jwarp.2014.62018.

CHAPTER 4

R. Burt, T. Weber, S. Park, S. Yochum, and R. Ferguson, "Trace Element Concentration and Speciation in Selected Mining-Contaminated Soils and Water in Willow Creek Floodplain, Colorado," Applied and Environmental Soil Science, vol. 2011, Article ID 237071, 20 pages, 2011. doi:10.1155/2011/237071.

CHAPTER 5

Saadia R. Tariq and Nabila Rashid, "Multivariate Analysis of Metal Levels in Paddy Soil, Rice Plants, and Rice Grains: A Case Study from Shakargarh, Pakistan," Journal of Chemistry, vol. 2013, Article ID 539251, 10 pages, 2013. doi:10.1155/2013/539251.

CHAPTER 6

A. O. Ojo, M. T. Adetunji, K. A. Okeleye, and C. O. Adejuyigbe, "Soil Fertility, Phosphorus Fractions, and Maize Yield as Affected by Poultry Manure and Single Superphosphate," International Scholarly Research Notices, vol. 2015, Article ID 616213, 8 pages, 2015. doi:10.1155/2015/616213.

CHAPTER 7

Amira M Embaby, Aliaa A Masoud, Heba S Marey, Nadia Z Shaban, and Tayssir M Ghonaim, Raw Agro-Industrial Orange Peel Waste as a Low Cost Effective Inducer for Alkaline Polygalacturonase Production from Bacillus Licheniformis SHG10, doi:10.1186/2193-1801-3-327.

CHAPTER 8

Berhanu Andualem and Amare Gessesse, Proximate Composition, Mineral Content and Antinutritional Factors of Brebra (*Millettia Ferruginea)* Seed Flour as Well as Physicochemical Characterization of Its Seed Oil, doi: 10.1186/2193-1801-3-298.

Index

A

Agricultural area 48
Agricultural production 40
Amino acid 198, 199, 215
Atomic absorption spectrometry (AAS) 71

B

Biodiesel production 193
Biomass production 2, 7
Biotechnological valorization 158

C

Carbohydrate content 192, 197
Cation Exchange Capacity (CEC) 8
Chemical Company 5, 14
Chemical compound 194
Chemical propertie 38, 40, 52
Citrus fruit 158
Colorado Department of Public Health and Environment (CDPHE) 61
Crop production 38, 39, 40, 48, 49, 51
Crude castor oil 203

D

Dinitrobenzoic acid [(DNSA) 180

E

Electrical conductivity (EC) 23, 71
Electrical Conductivity (EC) 38
Exchangeable Sodium Percentage (ESP) 38, 39

F

Flame Atomic Absorption Spectrophotometer (FAAS) 112
Food security 40, 51
Fractionation procedure 138

G

Groundwater 38, 39, 47, 48, 49, 51, 52

H

Hydrophobic nature 30

I

Irrigation water 37, 38, 39, 40, 42, 43, 48, 49

M

Middle Zone (MZ) 28
Mobility index value 113, 120, 129

N

National Institute of Standards and Technology (NIST) 74
NCBI [National Center for Biotechnology Information] 179
North Gaza (NZ) 27

O

One variable at a time approach (OVAT) 181
Optical Density (OD) 27
Organic matter 197, 198
Organic matter (OM) 71
Oxide bound (OX) 73

P

physicochemical parameter 112, 113, 114, 121
Phytoavailability of metal 123
Plant response 11
Polygalacturonase (PGase) 158
Poultry manure 135, 153
Poultry manure (PM) 135, 143, 144
Preventing eutrophication 151

Q

Quality control (QC) 74

R

Relative standard deviation (RSD) 74
Response surface methodology (RSM) 158
Rice-consuming countrie 111
River water 43, 49

S

Saline water 49
Six month 4
Slightly acidic soil 142, 153
Sodium Adsorption Ratio (SAR) 38
Sodium Adsorption Ratio [SAR) 39
Soil contamination 126
Soil organic matter 22, 29, 31

Soil sample 38, 42, 45, 47
Soil surface material 69
Statistical analysis system (SAS) 140
Subsequent mobilization 111
Super Absorbent Polyacrylate (SAP) 1
Super absorbent polymers (SAPs) 2

T

Total Exchangeable Bases (TEB) 42
Total extractable (TE) 72, 73

U

US Forest Service (USFS) 61

V

Vegetative tissues 110

W

Water quality 40
Water resource 11
Water sample 38, 42, 43, 52
Water saving 2
Water scarcity 2
Water soluble (WS) 73
Water stress 2, 3, 6, 7, 9, 10, 11, 12, 13
Wide range 3
Willow Creek Reclamation Committee (WCRC) 61